HOW TO MANAGE EXPERIENCE SHARING: FROM ORGANISATIONAL SURPRISES TO ORGANISATIONAL KNOWLEDGE

Related books

AHMED, LIM & LO	Learning through Knowledge Management
FRICK	Systematic Occupational Health and Safety Management
HALE, HOPKINS & KIRWAN (eds.)	Changing Regulation: Controlling Risks in Society
HALE & BARAM	Safety Management and the Challenge of Organisational Change
HALE, WILPERT & FREITAG	After the Event: From Accident to Organisational Learning
PRUSAK	Knowledge in Organisations
VINCENT & DE MOL	Safety in Medicine
WILPERT & FAHLBRUCH	System Safety: Challenges and Pitfalls of Intervention

www.elsevier.com

Related Journals

Accident Analysis and Prevention
Editor: F.A. Haight

Information Processing and Management
Editor: T. Saracevic

Journal of Operations Management
Editor: R. Hanfield

Journal of Safety Research
Editors: T.W. Planek and M.-L. Lin

Safety Science
Editor: A.R. Hale

HOW TO MANAGE EXPERIENCE SHARING: FROM ORGANISATIONAL SURPRISES TO ORGANISATIONAL KNOWLEDGE

Edited by

J. H. ERIK ANDRIESSEN
Delft University of Technology, Delft, The Netherlands

BABETTE FAHLBRUCH
Berlin University of Technology, Berlin, Germany

2004

ELSEVIER

Amsterdam – Boston – Heidelberg – London – New York – Oxford
Paris – San Diego – San Francisco – Singapore – Sydney – Tokyo

ELSEVIER B.V.
Sara Burgerhartstraat 25
P.O. Box 211, 1000 AE
Amsterdam, The Netherlands

ELSEVIER Inc.
525 B Street
Suite 1900, San Diego
CA 92101-4495, USA

ELSEVIER Ltd
The Boulevard
Langford Lane, Kidlington,
Oxford OX5 1GB, UK

ELSEVIER Ltd
84 Theobalds Road
London WC1X 8RR
UK

First edition 2004

Library of Congress Cataloging in Publication Data
A catalog record is available from the Library of Congress.

British Library Cataloguing in Publication Data
A catalogue record is available from the British Library.

ISBN: 0-08-044349-4

⊗ The paper used in this publication meets the requirements of ANSI/NISO Z39.48-1992 (Permanence of Paper).
Printed in The Netherlands.

Working together to grow
libraries in developing countries

www.elsevier.com | www.bookaid.org | www.sabre.org

ELSEVIER   BOOK AID
             International   Sabre Foundation

CONTENTS

FOREWORD

Bernhard Wilpert[1][2]

This volume is part of a long running scientific endeavour emerging from an international interdisciplinary study group on "New Technologies and Work – NeTWork" and presents its outcome from the 19th workshop in 2001.

NEW TECHNOLOGIES AND WORK

NeTWork is an international, interdisciplinary study group with the objective to advance the study and intellectual penetration of social and scientific problems posed by he diffusion of modern technologies in all domains of work life. Over two decades NeTWork, partially supported by he Werner Reimers Foundation (Bad Homburg, Germany) and the Maison des Sciences de l'Homme (Paris, France), has held annual workshops relating to the overall theme of new technologies and work. These have covered a wide range of topics that included human error, training, distributed decision making and management (see below).

While the original activities of NeTWork began with a wide coverage of sub-themes, more recent preoccupations focused more specifically on safety of high technology systems and the role of human contribution to either breakdowns or the maintenance of hazardous systems. In this volume however the scope is again broadened to an issue that is not only pertinent to the safety area but to modern organisations in general, i.e. learning from experiences and sharing what is learned with colleague. The motivation to deal with knowledge sharing was manifold.

---

[1] Berlin University of Technology, Germany
[2] This paper is based on a previous draft by J. Reason and B. Wilpert

The idea for this theme arose from the confluence of deliberations of the core group of the NeTWork group and some observations, which we made during recent research projects.

The past few years have seen a succession of major disasters afflicting a wide range of complex technologies: nuclear power plants, chemical installations, spacecraft, roll-on-roll-off ferries, commercial and military air craft, off-shore oil and gas platforms, telecommunication systems, computerized stock exchanges, high-tech medical services, hazardous good transport and modern railway networks.

At a purely surface level, these accidents varied widely in their specific causes and consequences. At a more fundamental level, however, they shared a number of important features:

(a)   They occurred within well-defended systems that were proof against single failure, either human or mechanical.
(b)   They arose from the adverse conjunction of several distinct causal events. Moreover, a large proportion of these multiple root causes were present within the system long before the accidents occurred.
(c)   Human rather than technical failures played a dominant role in all of these accidents.

A point has been reached in the development of modern organisations where the greatest dangers and the greatest successes are related to the capacity to develop and apply new experiences and solutions within the organizational, design and managerial sectors.

How NeTWork Works

The themes of annual workshops are planned and evaluated by a small 'core group' whose membership varies slightly according to the respective sub-theme to be treated. Two or three "godfathers" further detail a chosen sub-theme and propose a list of prospective participants. The choice of invitees is made internationally on the basis of their active research involvement with respect to the sub-theme or corresponding practice experience. Thus more than 210 persons from 21 countries participated in the NeTWork workshops between 1983 and 2003.

All contributions to a given workshop are distributed to all participants several weeks in advance in order to facilitate their thorough reading before the workshop itself. Only short statements are possible for each contributor summarizing the main points of his/her paper contribution in order to maximize thorough discussion and proposals for the improvement of each contribution. The workshop's "godfathers" usually serve as editors for the preparation of the publication which is usually developed from each meeting. So far 15 publications have appeared from these NeTWork activities (see below). Two more are in preparation. The model of this international and interdisciplinary exchange has meanwhile been adopted by various other institutions. The administration of NeTWork activities is managed by staff of the Berlin University of Technology: Prof. Bernhard Wilpert (Speaker) and Dr. Babette Fahlbruch (previously: Prof. Antonio Ruiz Quintanilla and Matthias Freitag, Workshop Coordinators). The Werner Reimers Foundation and the Maison des Sciences de l'Homme lend further administrative and material support.

THE PRESENT VOLUME

Looking at the themes of the previous years' workshops it seemed obvious that they constituted the historical setting and the roots of this volume. The workshop 1995 dealt with learning from events. The issue here was the feedback control of safety, i.e. here we looked at various ways people and organizations analyse their events and try to learn from their past operational experience, foremost from incidents or events. Another issue was also to identify the barriers to effective organisational learning. The contributions were published in A. Hale, B. Wilpert and M. Freitag (eds.) (1997). *After the event: From accident to organisational learning.* Amsterdam: Elsevier.

The 1996 workshop on "Safety management and the challenge of organisational change" gave a first attempt to look at the field in a more holistic approach, i.e. with the safety management perspective. The discussion focussed on how to model safety management systems and how to adapt to external change, i.e. intended and planned change and learning as well as reaction to external pressures. It resulted in a publication by A: Hale and M. Baram (eds.) (1998). *Safety management: The challenge of change.* Amsterdam: Elsevier.

In 1997 the workshop centred on the problem of "Coping with accelerating technologies". We tried to address the problem to maintain safe operations and avoid system breakdowns faced by a drastic technological change in some of the high hazard industries. A special issues in safety Science edited by Barry Kirwan (ed., 2001) became the publication outlet of our deliberations: Coping with Accelerating Socio-Technical Systems, *Safety Science, 37* (2-3).

The 1998 workshop was on "Risk and safety in medicine", which was a really new territory for the NeTWork group. The medical field was an area rather unfamiliar to the core group and traditional participants of the workshops. In consequence, the editing function was out-sourced to medical specialists: Charles Vincent and Bas de Mol, two internationally reputed colleagues in the field. The main question was concerned with possible transfer of learning in hazard industries to medicine. The publication from this workshop seems to become a classic of its kind in the domain: C. Vincent and B. de Mol (eds.) (2000). *Safety in Medicine*. Amsterdam: Elsevier.

In 1999 the workshop moved to the intervention field with "Achieving successful safety interventions". We tried to contribute to solve the problem when a safety intervention could be seen as successful, discussed the leverage points for implementation and how to sustain change or improvements. The contributions were published in B. Wilpert and B. Fahlbruch (eds.) (2002). *System Safety – Challenges and Pitfalls of Intervention*. Amsterdam: Elsevier.

The Millennium workshop in 2000 focused on " Safety Regulation: The Challenge of New Technologies and New Frontiers". We discussed the changing face of regulation in traditional risk industries as well as new domain for regulation. The result was a publication by B. Kirwan, A. Hale and A. Hopkins (eds.) (2002). *Changing Regulation: Controlling Risks in Society*. Amsterdam: Elsevier.

One of the conclusions of these reflections was that we seem now fairly well advanced in analysing safety problems. What appears to be missing was a better understanding of the methods and strategies, which could close the gap between lessons to be learned and the real learning in and of organisations. Therefore, an attempt was made to review the different approaches to learning and sharing of knowledge. Because knowledge management is an approach of nearly all kind of industries we widened our field from the limits of risky technologies to a wider range. This then formed the historical roots of the present volume.

As to the structure of this volume: This volume contains four main sections starting with basic concepts on knowledge management of and organisational learning from successes and errors with chapters by Andriessen and Fahlbruch (ch. 1), by Koornneef and Hale (ch. 2), by Huysman (ch. 3), by Dhondt (ch. 4), and by Lehner (ch. 5). The second section focuses on learning from errors with chapters by Kjellén (ch. 6), by Koornneef and Hale (ch. 7), by Marzi (ch. 8), by Wahlström (ch. 9), and by Wright (ch. 10). In the third section learning from successes is in the centre of the discussion in the chapters by Andriessen, Huis in't Veld, Poot and Soekijad (ch. 11), by Dhondt, Verbruggen, van Sloten and Kwakkelstein (ch. 12), by

Huysman (ch. 13) and by Maier and Lehner (ch. 14). The fourth section is meant to illustrate barriers and conditions of knowledge sharing with chapters by Sonnentag (ch. 15), by Hurley (ch. 16) and by Baram (ch. 17). This volume ends with a summary by Andriessen and Fahlbruch (ch. 18).

NETWORK PUBLICATIONS

Rasmussen, J., K. Duncan. and J. Leplat. (eds.), (1987). *New Technology and Human Error.* Chichester: Wiley.

de Keyser, V., T. Qvale., B. Wilpert and S. A. Ruiz Quintanilla. (eds.), (1988). *The Meaning of Working and Technological Options. Chichester: Wiley.*

Bainbridge, L. and S. A. Ruiz Quintanilla. (eds.), (1989). *Developing Skills with Information Technology.* Chichester: Wiley.

Warner, M., W. Wobbe. and P. Brödner. (eds.), (1989). *New Technology and Manufacturing Management. Strategic Choices for Flexible Production Systems.* Chichester: Wiley.

Rasmussen, J., B. Brehmer. and J. Leplat. (eds.), (1990). *Distributed Decision Making. Cognitive Models for Complex Work Environments.* Chichester: Wiley.

Andriessen, E. and R. A. Roe (eds.), (1994). *Telematics and Work.* Hove: Lawrence Erlbaum.

Wilpert, B. and T. Qvale. (eds.), (1993). *Reliability and Safety in Hazardous Work Systems - Approaches to Analysis and Design.* Hove: Lawrence Erlbaum.

Hale, A. R., M. Freitag. and B. Wilpert. (eds.), (1997). *After the Event - From Accident to Organizational Learning.* Amsterdam: Elsevier.

Hale, A. R. and M. Baram. (eds.), (1998). *Safety Management. The Challenge of Organisational Change.* Amsterdam: Elsevier.

Vincent, C. and B. de Mol. (eds.), (2000). *Safety in Medicine.* Amsterdam: Elsevier.

Wilpert, B. and B. Fahlbruch. (eds.) (2002). *System safety - Challenges and pitfalls of interventions.* Amsterdam: Elsevier.

Kirwan, B., A. R. Hale and A. Hopkins (eds.) (2002). *Changing Regulation: Controlling risks in society.* Amsterdam: Elsevier.

Andriessen, J.H.E. and B. Fahlbruch (eds.). *How to manage experience sharing: From organisational surprises to organisational knowledge.* Amsterdam: Elsevier.

Furthermore, based on NeTWork workshops three special issues have been published in learned journals:

de Montmollin, M. (ed.), (1990). Skills, Qualifications, Employment. *Applied Psychology*, *39* (2).
Leplat, J. and A. R. Hale (eds.), (1998). Safety Rules. *Safety Science, 29* (3):
Kirwan, B. (ed.) (2001). Safety and accelerating technologies. *Safety Science , 37 (*2-3).

PART 1

BASIC CONCEPTS

1

KNOWLEDGE MANAGEMENT OF SUCCESSES AND ERRORS

J. H. Erik Andriessen[1] and Babette Fahlbruch[2]

INTRODUCTION TO KNOWLEDGE MANAGEMENT

Since information and knowledge are becoming the central assets in information society, many organisations, both in service organisations and in traditional industries, may face questions of the following type.

- How can an organisation learn from its successes and failures, from its experiences and accidents?
- How can we prevent the loss of knowledge caused by intensive employee turnover or retirement?
- How can implicit knowledge be shared with colleagues and newcomers?
- To what extent can Information and Communication Technology (ICT) help to solve these problems?

Organizations like to learn from their mistakes and successes, they try to prevent the wheel to be reinvented anew, they want their employees to create new knowledge on the basis of experience. The problem has increased not only because the demand for information and knowledge has grown, but also because phenomena such as shorter product cycles, fierce

---

[1] Delft University of Technology; The Netherlands
[2] Berlin University of Technology, Germany

competition and job-hopping prevent organizations to rely on traditional means for knowledge transfer. In the German nuclear industry for example the first generation of employees since first nuclear power plants were put into operation is now about to retire. Because of this, management and staff in the nuclear industry are worrying about the loss of know-how and of know-why.

In the last ten to fifteen years a solution has been looked for in the development of large scale databases, expert systems, data-mining and other forms of 'knowledge technology'. The enormous potentials of ICT were expected to provide means for capturing, storing and distributing all the knowledge available. However, research studies have shown what many already suspected: the systems can help to store and make available large quantities of information. But information is not knowledge and availability does not guarantee actual use of what is available. Knowledge is information that is experienced and interpreted by humans, knowledge implies expectations and attitudes. Knowledge, moreover, can be explicit, but also be implicit, tacit. A project leader's knowledge concerning a project he has finished contains all the experience with the tools and the clients and the competition, why it took so long to finish, why he finally chose a certain strategy, how he succeeded in persuading the client, etc. This is the type of knowledge a colleague might want to hear if she is in a comparable position, and which the project leader is quite willing to tell her. But he loathes to put the whole story in a data-system, for many reasons: it takes too much time, he does not exactly know what to write, he certainly does not like to write down his mistakes into a large and impersonal database. And because he has found out that he himself has hardly benefited from the material others have put into the system.

The above-given analysis does not deny the value of information bases. Systematic inventories of accidents or near-misses have provided insights into the magnitude of certain safety threats or hazards. Systematic sales data have given insights into e.g. preferences of clients. But sharing knowledge concerning ways of working, preventing accidents, solving problems or dealing with clients requires not only the provision of bare information but also the interpersonal exchange of views and feelings.

This book has grown from a workshop that brought together researchers and practitioners from a wide range of areas, including the safety domain. The focus of the workshop was not on technical systems but on the question how organizations can learn, i.e. how knowledge can be shared and preserved in such a way that it can be used for preventing the wheel to be reinvented and for the creation of new knowledge. The participants came from widely different disciplines, such as safety consultancy, organisational psychology, computer science and law.

This accounted for wide differences in perspectives and of concepts used. The focus of some contributions was on individual and group processes, of others on organisational strategy or societal and environmental issues. The confrontation of these views however appeared to be very fruitful, and many common issues emerged from the seemingly different approaches.

It was discovered, however, that organisational knowledge management appears to be quite different when it is related to learning from failures in accident sensitive environments than when it is related to learning from success in service organisations. To illustrate the difference we present two case studies.

Two case studies

The Aviation Safety Reporting System (ASRS)
Since 1976 the National Aeronautics and Space Administration (NASA) runs the Aviation Safety Reporting System (ASRS) for civil aviation. All persons being somehow concerned with flying are asked to report about observable events, which are judged as safety relevant, such as dangerous situations, incidents or near misses. Reports can be delivered by pilots, cockpit-crews, air traffic controllers, ground level mechanics, pursers and travellers, i.e. by all people that are involved in aviation. All voluntarily reporting persons are assured of confidence as well as of immunity according to justice in the case of violations against rules, which were not committed on purpose. Immunity means here that self-reported information will not be used against the person if the violation was not intended, if it was not a criminal act, if the person did not deliver reports for which he asked for immunity within the last five years and if reporting took place within 10 days after the event.

From 1976 to 1992 more then 232.000 events were reported, of which nearly 50% were violations of rules. The reporting form consists of three parts: 1. Information about the reporting person(s), date and time as well location; this part will be deleted later for de-identification. 2. Information about task and function of the reporting person(s) during the event, their qualification and professional experience, information about the airplane, about the flight, about air traffic, and about external conditions like weather, sight. 3. A narrative report on the event.

The processing of reports implies an assessment whether the report is an event to be processed by NASA or a criminal act, attachment of reports of the same event from other persons, codification according to a codification scheme, obtainment of additional information by contacting the reporting person, de-identification of report, i.e. deleting the first part of the report and information within the other parts which could lead to the identification of the reporting person, quality control of de-identification, storing the codified report(s) into a data-base, storing the report(s) on microfiche, deletion of the original reports.

Each two weeks important events are presented and discussed with the regulatory body, the FAA, and important groups like pilot unions. On each report of dangerous situations either an Alert Bulletin or a For Your Information sheet is published, to distribute safety relevant information to concerned persons. Furthermore, there are monthly and quarterly publications. In each publication feedback is asked which is delivered in about 50% of the cases. Furthermore, the data-base is used for additional analyses and for research.

The 3D modelling network of BP

In BP the technique of 3D-reservoir modelling is used to improve the understanding of oil and gas fields, by building detailed computer models of the underground reservoir layers. The conviction is that a better reservoir description leads to better reservoir management and so to better reservoir performance (more oil, faster oil, less water etc.). These computer models are extremely detailed and complex, and may contain many million cells. 3D modelling is a fairly immature science. It only started to move out of the research centre into the business units in the early 1990's. Recently the software vendors have been producing software for the task, which anyone can use. This has broadened the use of the technique, but exposed the immaturity of the science, and resulted in a growing number of practitioners with a need to learn from each other. A conference in April 1996 highlighted the need for a network of reservoir modellers, and Ray King (an experienced geophysicist from Aberdeen) volunteered to organise the network. He based the network processes around an email discussion forum, initially on a list server, but later moving to MS Exchange public folders.

Ray recalls that "once we had it rolling in October 1996 it took off, and there was plenty of e-mail traffic. People got the hang of being able to reach over 40 technical professionals reliably, and get messages back. They can participate in the discussion, or can view the messages without committing to being on the list. The visible network, that is, those who are on the mailing list, now contains over 100 practitioners, some well-versed in 3d-reservoir modelling, others keen to learn. Many are practitioners in remote offices; places like Caracas or Kuwait". Currently the 3dmod head count stands at 98, in locations such as London, Aberdeen, Caracas, Houston, Anchorage, Kuwait and Bogota. There is also an invisible network, people who access the public folder and participate passively: They may decide to join in visibly, or not. If they do, they subscribe voluntarily by sending a note to Ray, and he puts them on the mailing list. For a year or two, the knowledge generated by the 3D Modelling network resided, unsorted, in the many hundred of messages in the public folder. Ray and the network realised that this was not the ideal solution; knowledge was becoming hard to find, and the same questions were being raised time and again. The 3D modellers network have now set up an Intranet site where they store and share distilled knowledge, frequently asked questions, and best practice. The network receives no funding from the central organisation, but involves very little expense. Ray King spends about 5% of his time organising the network, which is largely self-sustaining. The members contribute their own time and expertise to the discussions, and in return gain timesaving tips and solutions. The network holds no performance contract, and delivers value by increasing the efficiency of the members, so they can build complex 3D models in rapid time.

Differences and commonalities

From the above two cases it becomes clear that we are dealing with two widely different situations. These differences will be discussed in section 1.4. Nevertheless, in both cases we deal with common phenomena, i.e. with the collection, exchanging, and conveying of highly relevant *information* and *knowledge*. In both cases the people involved are concerned with systematic management of this knowledge both in databases, i.e. *codification*, and via interpersonal interaction, i.e. *personalisation*. All these processes are part of explicit or implicit strategies of *organisational learning*, where *knowledge transfer* mechanisms are used to develop *organisational memory*. In the following section these central and common concepts will be clarified and discussed.

Central concepts

This part chapter serves to clarify the central theories and concepts. Some will be dealt directly in this chapter, others will be dealt within the four short chapters hereafter.

Knowledge versus information:

The traditional distinction is that between data, information and knowledge. Data are simple, objective givens, such as figures or facts about events. Information is data in a context which gives the data a meaning. Knowledge is information that is experienced by a person in a particular situation and is therefore connected to feelings and sense making. Knowledge is contextualised information. Strictly speaking when knowledge is explicated into a data-system it is no longer contextualised and personally sense making, so it became information again. But the difference between information and knowledge is not sharp, it has rather a scale with grey zones: some information is hardly more than data, such as lists of names, and some information is almost knowledge, such as the information in an expert system.

On this scale from information to knowledge we can also place the concepts of *explicit and implicit or tacit knowledge*. Explicit knowledge is has been expressed in words, it is codified and structured into e.g. documents, systems and manuals. Tacit knowledge is highly implicit personal knowledge, such as experiences, intuitions, hunches. Organisations are constantly trying to make knowledge explicit, so that it can be transferred and stored easily, according to a *'codification* strategy'. Tacit knowledge is difficult to exchange and store systematically. It is much better be exchanged through a *'personalisation'* strategy (see below).

Knowledge Management:
Knowledge management is the management of knowledge processes. Various types of knowledge processes have been distinguished in the literature such as developing, storing, distributing, applying, evaluating knowledge (the 'knowledge value chain of Weggeman, 1997), or process of knowledge management (Probst, Raub and Romhardt., 1999) Managing these processes implies systematic activities with the aim of an optimal functioning of these processes. It is still strongly debated to which extent this requires directive guidance and supervision or only a kind of facilitation, i.e. the provision of favourable conditions (see e.g. Huysman in this Volume). Knowledge management might better been called knowledge process organising, to prevent associations of authoritarian pushing and control. The term knowledge management has however become so common place that we have to do with it.

Codification versus personalisation approach
The **codification strategy** for knowledge management (Hansen *et al.*, 1999). implies a focus on procedures to elicit tacit knowledge from employees, converting it into explicit knowledge and storing it in company wide repositories. This approach has seen many failures, mainly because much knowledge – in the sense of personal experiences, based on values and attitudes – is very difficult to explicate, and also because of the psychological resistance against providing and against using this *impersonalised* knowledge.

A second way of dealing with sharing, applying and developing knowledge is what Hansen *et al.* call the *'personalisation strategy'*. In this strategy the focus is on people meeting each other, on interpersonal knowledge sharing in master-apprenticeship relations and on what is called 'communities of practice' (see Andriessen *et al.* in this volume) . The role of ICT applications is limited to supporting the communication between people and to e.g. 'yellow pages', i.e. a database containing information on which experts to find where. The authors convincingly argue that the first strategy is called for in organisations (or departments) where the information processes are rather routinised, e.g. consultancies with more or less standard projects. The personalisation strategy is appropriate for organisations (or departments) with non-standardised, i.e. creative, novel production processes.

Learning from failure and learning from success

Is learning from failure different than learning from success? Is knowledge management in the first case different from that in the second case? That is one of the central issues in this book. Learning from failure is mainly organised in high hazard industries, because in these organisations it is of high importance not to be confronted with the same or comparable incidents or accidents again. The learning process is here sometimes even regulated by the government or by external agencies or regulatory bodies. The German nuclear industry for instance has various instruments to learn lessons from operational experiences. Systematic event analyses combined with reporting systems as well as institutionalised feedback loops serve to promote the sharing of knowledge. As example could be given the system of event reports to be circulated. Here an external agency re-analyses world-wide event reports from the nuclear industry. Significant events are selected and extended reports are formulated. These reports are distributed to each German power plant together with questions to be answered like "Is an event like this possible to happen in your plant?" "What kind of measures will you take that this will not happen to your plant?" "Although this plant my vary in type of technology or type of organisation, could there be lessons to be learnt for your plant as well?" The plants are asked to discuss the problem and write a report with their comments and answers to the questions.

In sum the feature of learning from failure could described in the following terms:

- Very formalized approach
- Systematic analysis of failures
- Various methods for complex analysis
- Codified reports
- Institutionalised feedback loops
- High value for information gained
- Changes of processes and methods are risky

Learning from success can be separated in two different approaches: Learning from success in the past in the sense of best practices and learning with an eye on generating new knowledge, i.e. innovative knowledge sharing.

The learning from successful past experience is a typical feature of service orientation, such as in quality management of standard industries, or in consulting companies. Here organisation members are asked to store or distribute their experiences and insights concerning e.g. project performance or other issues. This exchange of experiences and insights may become systematised in the form of best practices, that means rules for future action related to more or less specific situations. In sum this context is characterised by the following aspects:

- Less formalization
- Less systematic analysis of experience
- Less methods for complex analysis
- Reports are less formalised, but still written
- Less institutionalised feedback loops
- Value for information gained is lower
- Changes of processes and methods are less risky

Innovative knowledge sharing requires the generating of new knowledge going beyond the existing one. Here it seems of high importance to build on personal exchange instead of codified material . Because of that formalisation is often seen as a barrier, there is no systematic approach to analyse past experience, often there are no reports and no institutionalised feedback loops, but changes are seen as very important as an opportunity for the future.

The discussion on these two types of learning and the implications for knowledge management will be taken up again in the last chapter of this book.

REFERENCES

Hansen, M.T., N. Nohria. and T. Tierney (1999). What's your strategy for managing knowledge. *Harvard Business Review,* **77**, 106-116.

Probst, G. J., S. Raub and K. Romhardt. (1999). *Wissen managen. Wie Unternehmen ihre wertvollste Ressource optimal nutzen* (3rd edition). Gabler, Wiesbaden.

Weggeman, M. (1997). *Kennismanagement: Inrichting en besturing van kennisintensieve organisaties.* Scriptum, Schiedam, The Netherlands

2

ORGANISATIONAL LEARNING AND THEORIES OF ACTION

Floor Koornneef and Andrew Hale[1]

INTRODUCTION

Organisational learning is an essentially human process: an organisation learns through people. A basic notion in psychology is that *learning* is the process of solving a problem that cannot be avoided. If the successful approach (= lesson) is not retained for reuse in future, learning must be repeated when the problem recurs. However, the effectiveness of learning remains low when lessons are retained but cannot be retrieved for use in a timely way. In order to learn as an organisation, the organisation must organise its learning!

To be a "learning organisation" has been one of the highest ideals held up for companies to achieve in the last decade (Senge, 1990). In the field of safety and health management we have been trying to set up systems in companies and at national level to learn from accidents and incidents for over 15 years. However, despite these efforts, there are far more failures than success stories (Schaaf *et al.*, 1991; Hale *et al.*, 1997). In a recent study Koornneef (2000) reviewed ten case studies on attempted introductions of such learning systems. One of his main conclusions is that the difficulty of setting up and sustaining these learning systems has been seriously underestimated. Organisational learning requires organisation and investment of time and resources. Above all it must be anchored firmly at the level of the primary processes of the

---

[1] Delft University of Technology

organisation (the shop floor, the coal face, the operating theatre, the control console, etc.), so that it involves and rewards the efforts of people working there. If those individuals cannot learn from the system, and do not see the value of the learning process, the system will never work.

This chapter describes the essentials of a learning system and the barriers to learning which have to be overcome.

ORGANISATIONAL LEARNING FRAMEWORK

Argyris and Schön (1974, 1996) have researched how organisations learn effectively. Argyris built his systems approach on basic cybernetics as presented by Ashby (1960, 1956), on the work of Bateson (1972) regarding learning, and on that of Dewey (1938) about 'inquiry'.

> *"The term learning either means a product of the learning process – 'something learned – or the process that yields such a product... An organisation may be said to learn when it acquires information (knowledge, understanding, know-how, techniques or practices) of any kind and by whatever means. ...all organisations learn, for good or ill, whenever they add to their store of information, and there is no stricture on how the addition may occur. The generic schema of organisational learning includes some informational content, a learning product; a learning process which consists in acquiring, processing, and storing information; and a learner to whom the learning process is attributed." (Argyris and Schön, 1996, p. 3).*

It may be said that the organisation will learn

> *"when its members learn for it, carrying out on its behalf a process of inquiry that results in a learning product... Inquiry does not become organisational unless undertaken by individuals who function as agents of an organisation according to its prevailing roles and rules." (Argyris and Schön, 1996, p. 11).*

The term 'inquiry' is used here in a more fundamental sense of:

> *"the intertwining of thought and action that proceeds from doubt to the resolution of doubt. In Deweyan inquiry, doubt is construed as the experience of a 'problematic situation', triggered by a mismatch between the expected results of action and the results actually achieved." (Argyris and Schön, 1996, p. 11).*

The output of such 'organisational inquiry' takes the form of a change in thinking and acting that yields a change in the design of organisational practices. Such *knowledge* can be stored in a way that is accessible, or can be assimilated into custom and practice.

THEORIES OF ACTION

Organisational knowledge about how to realise the organisation's objectives are represented through so-called '*theories-of-action*'. These include strategies of action, the values that govern decisions and the assumptions on which they are based.

A theory-of-action takes two different forms, that of an '*espoused theory*' that is explicitly advanced to explain or justify a given pattern of activity, and, simultaneously, that of a '*theory-in-use*' which is manifest only in action and may not be articulated or articulable. 'Espoused theory' is explicit and overt and 'theory-in-use' is implicit and tacit.

The organisation writes down 'espoused theories' in its manuals and when it describes how something should be done. It is 'espoused theory' when an organisation states that it runs business processes on a zero-accident basis. However, the 'theories-in-use' reflect actual practice as conceived by the operative persons. They envisage how things are done best: and act in accordance with their individual 'theory-in-use' of running processes. Members of a successful team have highly similar or matching theories-in-use. It is ideal if 'espoused theory' is matched to the accompanying 'theories-in-use' so that the actual organisational operations fit the proclaimed, intended actions.

Schein (1992) makes a similar distinction in his analysis of organisational culture. What matters in organisational learning is that the theory-in-use is changed. The espoused theory should change too so that there is consistency between documentation and practice. However, it is all too easy to change espoused theory, by rewriting procedures, without changing practice at all; safety rules are notoriously subject to lip-service and not well followed.

The tacit theories-in-use, including norms and values regarding operational performance, are manifested in behaviour within the organisation. On the other hand, an espoused theory may have nothing to do with the way in which the organisation actually operates. It may be no more than the philosophy that is trotted out when someone from outside poses questions, or something which the organisation would like others to believe that they do.

DOMAIN OF ACTION – OR WHAT CAN BE LEARNED

Learning is about changes that attract our attention. Some changes are disturbances that may require control actions. Good causal models help to get a better understanding of the dynamics of the processes that generate the disturbance. Espejo *et al.* (1996) contend that such models need to be grounded in shared views about the 'action domain' if they are to support organisational learning.

One cannot learn from operational disturbances without knowing what processes are being looked into: Espejo's action domain. Interpretation of disturbance data requires context knowledge about what processes are run within which organisational settings. The variety in actual disturbances and their contexts is large, so, already at the notification and analysis stage of learning we need a way of generalising the specific messages from the disturbed process. The 'Nertney wheel' (Bullock, 1979; Nertney, 1987) provides a general way of dealing with this. The basic ingredients of work, which constitute the organisation's operations, are People, the means including technology they use ("Plant"), and the objectives, organisational structures as well as rules on how to realise the goals ("Procedures"). To assure operational readiness the PPP-ingredients of work need to be balanced with respect to each other; this is symbolised by the "bull's eye" in Figure 2.1. The interfaces between these ingredients are therefore equally important: people and procedures, people and hardware, and procedures and hardware should all match each other. For instance, the instructions for an operator should be in a language that the operator understands.

An incident reveals an imbalance between the ingredients of work and the interface domains, shaded parts in Figure 2.1, which can be symbolised by saying that the "bull's eye" has moved off-centre. For instance, if we assume some erroneous behaviour by an operator, the system is off target, above the centre in Figure 2.1. In order to get the processes right again, one can get on target again by investing in the operator to improve behaviour. Alternatively, we could restore readiness through reallocation of tasks, e.g. to hardware, and modifying a protocol in order to improve the fault-tolerance of the whole system.

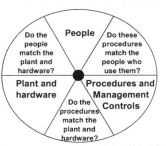

Figure 2.1: Ingredients of work: Nertney Wheel [simplified from (Bullock, 1979)]

Using Nertney's Wheel, inquiry can focus on the basic ingredients of work processes, their configuration and interaction, along with the underlying design and development of each component and its interactions. Lessons developed this way, can be communicated back to the shop floor process using the Wheel's structure to develop detailed adjustments that can be adopted into the organisational unit 's theory-of-action.

We therefore focus on disturbances or *operational surprises* in processes that an organisation runs or is involved with to realise its objectives. Our prime focus for learning is on the organisation's competence to manage risks that are inherent in its operations or process outputs that can jeopardise the realisation of these objectives. Thus, the lessons that can be learned by the organisation concern the way the organisation has allocated and configured resources and activities to meet its goals.

THE ORGANISATIONAL LEARNING PROCESS

Argyris (1992) emphasises that learning should be embedded in the whole organisation as a part of its normal operation. It is not an add-on extra. This means, in terms of safety, that there must be an intimate link between the risk assessment process, which specifies what hazard scenarios there are, the management process, which establishes control strategies and practices for them, the operational process which carries them out and the learning process, which evaluates, improves and fine tunes these controls.

The goals of the organisation are realised through processes run by organisational units. Learning starts in such an organisational unit that is performing activities for which operative members already have a "theory-in-use" in the form of routines, priorities and actions. These all set up expectations about how the activity will proceed and what consequences will arise from the actions taken. The activities are conducted with resources made available to the unit and under objectives, values and means, which are also given to or imposed on the unit. Basic resources are knowledge, technology, organisational structures, norms and rules, as well as people who deploy these resources in working processes to realise the organisation's objectives.

One of the normal expectations is that accidents will not happen. In most situations reality will match expectations. Sometimes, however, an operational surprise occurs in the form of an unexpected outcome. For health and safety management we are interested in the surprises which lead closer to danger. If the individual detects the surprise and changes the way of working as a result, individual learning has taken place. However, for organisational learning to take place, the individual must notify a relevant learning agent. This process of notification needs to have as low a threshold as possible, i.e. form no obstacle for the individual, so that it takes little time and effort. Such obstacles are, for instance, fear of blame, undue administrative burden in addition to job workload, and experiences of hearing nothing from previous notifications. Notifiers should not have to worry about classification of the reported operational surprise, e.g. whether or not it is reportable. They should simply be encouraged to report anything of interest - if in doubt, report. The handling of insignificant events or conditions can be solved within the organisational learning process that is triggered by the notification (Koornneef, 2000) This is addressed in Chapter 7.

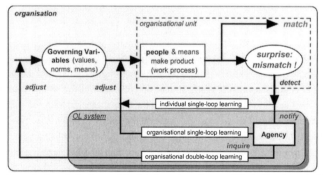

Figure 2.2: Organisational single- and double loop learning modified after Argyris (Koornneef, 2000)

Figure 2.2 indicates that learning has not occurred until a new match or a mismatch is produced. Therefore, learning cannot be said to occur at the point where someone (acting for the organisation) discovers a new problem or even when he or she invents a solution to that problem. By Argyris' definition, learning occurs only when the invented solution is actually produced (Argyris, 1982, 1992) and has been implemented. Learning includes the effective implementation of solutions to the problem encountered. The learning process 'ends' when the process outcomes match the expectations according to that part of the adjusted theory-in-use when it is next invoked.

Single- and double loop learning

Argyris uses Ashby's definitions of single and double-loop learning (Ashby, 1960). Ashby distinguished between the adaptive behaviour of a stable system (in which all essential variables lie within their normal limits) from the changes to the definition of normal itself. For example, the normal operation of a thermostat is qualitatively different from changing its temperature set point.

Single-loop learning affects strategies of action and underlying assumptions of theories-of-action. That is, it affects the way operational goals are achieved without changing goals or values themselves. *Organisational single-loop learning* products are visible in the organisation's theories of action, e.g. as minor modifications in a task protocol. In contrast, *double-loop organisational learning* affects norms, values and organisational targets that govern the organisational unit and its theory-of-action. Such changes actually mean modifications to the constraints for operations run by the unit, requiring adjustment of its theory-of-action and, thus, of the prevailing theories-in-use (Koornneef, 2000). The organisation's function is

> *"...to decompose double-loop learning issues into single-loop issues, because they are then more easily programmable and manageable. Single-loop learning is appropriate for the routine, the repetitive issue – it helps to get the everyday job done. Double-loop learning is more relevant for the complex, non-programmable issues – it assures that there will be another day in the future of the organisation." (Argyris, 1992, p. 9).*

Individual single-loop learning occurs when an individual detects an operational surprise in the organisational unit and, therefore, must decide what to do - take action or not, notify or not. He or she may take immediate corrective action and the lessons learned this way may then modify the individual's version of the applicable theory-in-use. Individual single-loop learning may be effective for the individual's operational actions, but does not count within the organisational unit of which the individual is a member: without notification, learning products produced by the individual are not adopted in the organisation's theories of action.

Notification is a prerequisite for organisational learning from instances of single 'surprise' experiences by individual members of the organisation. In this context, Bateson (1972) identified the phenomenon of 'Zero learning' which takes place when a message is received by an individual: the option to change the context that existed before receipt of the message has become apparent. This individual experience of detection of an unanticipated change in events or conditions is essential for learning: without detection of an operational surprise, there is nothing to notify to a learning agency.

Learning agency

Organisational learning from individual surprises is based on inquiry by agents assigned to learn for the organisation (Argyris and Schön, 1996). These agents may need to be grouped as a 'learning agency' (Koornneef, 2000).

Both organisational single- and double-loop learning requires a 'designated' learning agency, i.e. people who learn on behalf of the organisation, to ensure that the learning experience becomes embedded in the organisation. The learning agency should have the task of making additional inquiries if the notification is not enough to act as the basis for a decision.

There is a fundamental problem inherent in the notification process which is that the notifying message becomes divorced from its context in the process of communication. This complicates the process of learning by introducing variability in interpretation. Therefore, the learning agency has a role in recapturing and preserving the contextual information lost in the notification process. This is discussed further in chapter 7.

Lack of context in the notification can be compensated by the learning agency either by performing an in-depth inquiry or by its agents if they are very familiar with daily practice. In the latter case, the knowledge of the prevailing theories-in-use is largely available as the combined operational expertise of members of the agency (Koornneef, 2000). Thus, due to the variety of the operational contexts, which require to be known, a learning agency typically consists of more than one learning agent. The agency reviews an incident and generates lessons to be implemented directly by the organisational unit (single loop) or indirectly by changing values, norms and means for the functioning of the organisational unit (double loop). The lesson is actually learned when implementation prevents the incident (surprise) from recurring.

If the organisational unit can make the necessary changes to its working practices within its own resources and authority, "single loop learning" is sufficient to translate the lesson into new theory in use. A change in objectives and values form definitive marks of "double loop learning". A decision to change the process or to impose new values and norms, e.g. on performance or sustainability, calls for double-loop learning. The need to invoke a more senior person to authorise expenditure on equipment may still be single loop. Only sometimes does a change in management level indicate a change in the order of the learning loop. Only when the actual work practices have changed can we talk of organisational learning having taken place. As part of this process, the espoused theories need to be changed by updating hardware records, procedures, training course materials, policy documents, etc.

The learning process conducted by the learning agency in single or double loop mode results in a selection of options for change available, which must be conveyed to the organisational unit for choice and action, so that the unit adjusts its theory-in-use as proposed, see Figure 2.2.

Lesson learned

A lesson is not learned until the operative persons in the organisational units concerned adopt it in their mental model of the operation that generated the surprise. The effectiveness of the implemented changes must be monitored in order to secure closure of the (organisational) learning loops. A second mismatch related to the lesson contains the message that the selected options for change were either inadequate or incorrectly implemented and an additional cycle of organisational learning is imperative.

So far, three types of messages have been identified in a learning cycle that are essential in organisational learning processes:

- a 'surprise' message from the process to the observing member;
- the surprise notification message from the observer to a designated learning agency; and
- the 'lesson-to-learn'-message from the learning agency back into the organisational unit.

The first two messages have primarily an attention-grabbing function - an unexpected situation has occurred. This is different from the third message which add value through the operation of the learning agency. The third message must be understood and adopted by the organisational unit in order to complete the learning process cycle (Argyris, 1992, 1996).

Organisational learning requires that memory can be accessed from within the organisation. This is needed to share lessons learned. One key argument is that the learning agency should not spend time and resources on learning the same lesson twice from recurring incidents given a particular type of incident and its operational context. Moreover, lessons may have to be taught more than once, e.g. to people who are new in the process or function in a similar, but different unit.

COMMUNICATION IN ORGANISATIONAL LEARNING PROCESSES

Individuals differ in their sensitivity to and perception of operational surprises (in our case those related to risks). This variation may be explained in terms of (safety) culture. The safety culture can be expressed through prevailing theories-in-use; manifest in both risk perception and the sense of empowerment to act on or notify any problems affecting safety performance (Westrum, 1991; Guldenmund, 2000). If the safety culture is poor, individuals will have great difficulty to detect and recognise an emerging operational risk, or may be disinclined to notify. Strong organisational defences (Argyris and Schön, 1996) may be triggered where there are conflicts between theory-in-use and espoused theory or within espoused theory. A stated policy of open-minded safety culture - where people are encouraged to communicate freely about safety problems encountered - conflicts seriously with a practice where individuals who have had two accidents in a year, are cautioned and threatened with discipline. This conflict will block reporting and productive organisational learning, because individuals who experience an operational surprise will make up their minds to keep quiet about the surprise. Different strategies and provisions can be thought of to improve such a safety culture. The most essential is to reward incident reporting and restrict disciplinary action to blatant breaches of agreed duties of care.

For surprise notification, the threshold for signalling to the organisational learning system that an operational surprise experience has occurred must be made as low as possible so that the act of signalling becomes an automatic part of the prevailing theories-in-use within the organisational unit. The signal must always lead to follow-up action of one sort or another to close the loop back to the notifier. A high threshold might relate to the need to judge an event e.g. in order to select the most appropriate notification procedure out of many, but also to the lack of useful feedback to notifying persons and their organisational units or to a prevailing "blame" culture.

The subsequent assessment and treatment of the surprise notification involves processes of filtering and abstraction, but also of reintroduction of context and variety. The former is needed to generalise data for decision and to avoid overloading communication channels (feedback lines in Figure 2.2). The latter is needed to give decisions context and make them suitable and understandable or credible for implementation. Communication channels and their interfaces need to be designed for this purpose.

Learning how the organisation is learning is yet another important type of double-loop learning. This 'deuterolearning' (Bateson, 1972) is indicated when organisational single- and double-loop learning do not result in adequate 'lessons': in which case the configuration of the organisational learning processes, the composition of learning agencies, methods used for inquiry, or communication channels deployed may need revision.

THE KEYHOLE VIEW: PITFALL OF NOTIFICATIONS

Any notification system collects data in conformity with a predefined data model (form). This imposes a *'keyhole' view* of reality. Depending on the purpose of the notification data, completeness may be less or more important. The message may just be an 'alert' signal requiring little detail, or may be the input to a review process where the description on the form is crucial for problem diagnosis. Any report, by its choice of words simplifies real-life processes and conditions. It is impossible to know just on the basis of the data provided in the notification report whether or not crucial context data are missing and if so, which data. It is like looking into a room through a keyhole: one cannot see what is out of sight, but possibly relevant for understanding the visible scenes. There are ways to overcome this fundamental problem, but not by expanding the data model (Koornneef, 2000). That will only overload the notification system and discourage the reporting of surprises. One solution is to assign a learning agency appropriately embedded in daily operations: i.e. somebody who (knows what) is "in the room". Another is to limit the objective of the signal to that of alerting a relevant body that something unexpected has happened which might need a closer look: the message might be nothing more than "Surprise! Here and now!" This learning agency can then send someone "into the room" to have a closer look.

Throughput Modelling

It is sufficient for organisational inquiry to describe only the deviations from the expected state changes if the condition is met that the prevailing set of theories-in-use at the operational level are 'known' to inquiry agents, i.e. they know, in a particular situation **S,** that a desired consequence **C** results from a specific action **A** ($A \rightarrow C \mid S$). This condition can be met by modelling the prime operations of the organisation into meaningful operational steps that are distinctive regarding task execution, resources and, therefore also, are governed by distinctive sets of theories of action. The notification data then must include a reference to the appropriate process stage. See Figure 2.3 for an example.

A particular throughput phase is usually assigned to one organisational unit. The transition from one stage to the next one is typically based on a go/no go decision by the initiating (receiving or sending) unit.

The throughput stages of relevant hospital processes can be modelled from the admission to the discharge of individual patients according to a product life cycle. Peculiar in health care is that individual patients form the product. The means of adding value to the product during the process is by delivering care, diagnostic services and treatment. Once a patient has been admitted for treatment and care, he is supposed to come out again in an improved state of health or quality of life: product quality control practice does not allow the rejection of that patient before discharge (unlike in manufacturing industry).

The phases and embedded critical phase-steps as well as the ownership transitions provide concrete links for domain experts to process contexts relevant to the reported operational surprises. In other words, simply indicating which phase a surprise occurs in triggers rich contextual knowledge by an expert, who can then recognise and understand very brief descriptions of the surprise, without the need for detailed description.

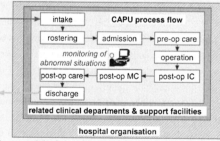

Figure 2.3: Throughput model of a <u>cardiopul</u>monary surgery department as 'seen' from the point of view of a CAPU patient. The monitoring of abnormal situations covers all stages. Causal factors underlying an operational surprise in one stage may originate from preceding stages where implementation of measures to resolve the causal factor might be easier than in the stage where the surprise occurred.

The notification of an operational surprise regarding organisational functioning is in essence a signal of a deviation from expectations. The important part of the notification is the depiction of the identified operational deviation. If we were to require that a notification must also contain a description of the applicable theory-in-use which was 'violated', we would create a serious motivational problem. Not the least problem is that, more often than not, relevant theories-in-use are tacit and sometimes cannot even be deduced by observations, see (Argyris and Schön, 1996, pp. 12-15). Hence, the individual must work to make them explicit and describe them. The need to describe the violated theory-in-use is avoided by coding the report to an explicit throughput process step, see text box above, which refers to learning in the cardiopulmonary surgery unit of a hospital (Koornneef, 2000). Experts in the system know from the code what to expect at this step. If the information is later to be used by, for instance, external non-experts, the theory-in-use can be reconstructed by interrogating domain experts in retrospect.

The keyhole view problem is fundamental as any free and coded data about the real world represents a model whose validity criteria and limitations are not defined in the case of surprise notifications. The solutions described above fit the needs of variety attenuation and variety amplification as identified in cybernetics approaches (Espejo *et al.*, 1996; Beer, 1979; Ashby, 1956).

BARRIERS IN ORGANISATIONAL LEARNING

Argyris and Schön (Argyris and Schön, 1996) have identified a characteristic style adopted by people dealing with threatening problems or situations in organisations. This style seems to be an almost automatic, defensive reflex of which the person seems to be unaware. During their research, they have refined the description of this response, which they refer to as "Model I theory-in-use". It is characterised by four rules-of-thumb and two strategies:

Rules-of-thumb:
- Strive to be in unilateral control
- Minimise losing and maximise winning
- Minimise the expression of negative feelings
- Be rational

Strategies
- Advocate views without encouraging inquiry
- Unilaterally save face (own and others)

Model I theory-in-use behaviour hinders change, especially when individuals behave defensively, for example when facing embarrassment or threats. Productive organisational learning requires a double-loop process and, thus, a shift from Model I to Model II behaviour, also to resolve errors that blindfold the learning agency in single-loop learning (Argyris and Schön, 1996). Model II theory-in-use behaviour is characterised by openness and mutual respect, also when questioning prevailing theories-in-use. Governing variables of Model II theory-in-use are valid information, informed choice and responsibility to monitor how well the choice is implemented in order to detect and correct errors, thus, inviting more learning. Organisations stuck completely in Model I behaviour will have great difficulty in learning, especially from unwanted situations, such as incidents which could lead to harm and hence liability. Model I symptoms of obstacles to organisational learning include

- unsupported attributions and evaluations e.g. "You seem unmotivated"
- advocating courses of action, which discourage inquiry e.g. "Lets not talk about the past: that's over!"
- treating ones' own views as obviously correct
- making covert attributions and evaluations
- face-saving moves such as leaving potentially embarrassing facts unstated

Kingston (Kingston, 2001) summarises twenty obstacles (Table 2.1) and ten ways to avoid organisational learning (Table 2.2) as characteristic of Model I:

Table 2.1: Twenty obstacles to learning
- perceived lack of time
- blame culture
- resistance to change
- lack of accountability
- wrong sort of accountability
- few opportunities for lateral communication
- too much top-down management
- "not my problem"
- sometimes: "shoot the messenger"
- passive communication (i.e. do not require change in behaviour or beliefs)

- tendency to "dig the detail"
- impersonal styles of communication
- alienation
- poor quality of relationships
- not enough "time-out" to talk
- poor commitment to lifelong learning
- specialists "own the message"
- specialists "own the problem"
- tightly comfort zone (e.g. only engineering issues are thought legitimate to consider)
- lack of trust

Table 2.2: ten ways to avoid learning
- do not collect or preserve information
- believe that "error" is a sufficient explanation
- do not use appropriate methods of analysis
- ensure that only specialists do investigations
- do not debrief at each level of line, especially operational personnel
- rely on formal communication alone
- ensure that solutions are the sole output of investigations
- do not own or track remedial processes
- only debate the technical details
- do not question your methods and motivations

Starting from Argyris' model of learning loops, Espejo *et al.* link incomplete organisational learning cycles with seven obstacles to organisational learning, four in single- and three in double-loop learning (Espejo *et al.* 1996, pp. 154-176), see Table 2.3 and Table 2.4 below:

Table 2.3: obstacles to single-loop learning
- *role-constrained* and *audience-restricted* learning: individual conceptual learning takes place, but operational learning is inhibited
- *superstitious* and *ambiguous* learning: operational learning takes place in an individual or organisational context, but conceptual learning is hindered

Table 2.4: obstacles to double-loop learning
- *superficial* learning: individual double-loop learning is inhibited
- *fragmented* and *opportunistic* learning inhibit organisational double-loop learning

Productive double-loop learning requires Model II behaviour and needs to be organised, since barriers to organisational learning must be overcome.

CONCLUSION

The principles of organisational learning as described by Argyris pinpoint the need to actively organise the learning processes of an organisation that wants to learn. Learning does not take place unless the learning process loop is closed. An organisation may learn through the people who work for it, but only if one or more of the members are assigned to be a member of a learning agency to learn for it. Single- and double-loop learning starts with 'zero learning' (Bateson, 1972) when an individual notices an unexpected operational condition ('surprise!') and may or may not inform the learning agency. The prevailing culture within the organisation has a strong impact on 'zero learning'. It may inhibit the recognition of a surprise condition, and subsequent notification. Once reported, the organisational unit where the surprise occurs may be able to resolve the surprise condition within its regulatory limits (single loop learning). If it cannot, higher management must decide about changing objectives, norms and values under which the unit functions (double-loop learning). The concept of 'theory of action' (Argyris and Schön, 1996) provides insight into the limited impact of the explicit 'espoused' theories about how things should be done compared to the tacit 'theories-in-use' about how things are really done. In defensive organisations the theory-in-use is in conflict with the prevailing 'espoused' theory and there is no way to put this difference on the learning agenda. This concept of 'theory of action' emphasises the limitations of top-down approaches when setting up organisational learning, as only the 'espoused' theories will be propagated.

REFERENCES

Argyris, C. (1982). How learning and reasoning processes affect organisational change. In: *Change in organisations: New perspectives on theory, research and practice* (P.S. Goodman and Associates, eds.), Jossey-Bass, San Francisco, USA.

Argyris, C. (1992). *On organizational learning*. Blackwell, Cambridge/Oxford, UK.

Argyris, C. and D. A. Schön (1974). *Theory in practice: Increasing professional effectiveness.* Jossey-Bass, San Francisco, USA.

Argyris, C. and D. A. Schön (1996). *Organizational learning II; theory, method, and practice.* Addison-Wesley, Amsterdam.

Ashby, W. R. (1956). *An introduction to cybernetics.* Methuen, London.
http:\\pcp.vub.ac.be/books/IntroCyb.pdf

Ashby, W. R. (1960). *Design for a brain.* John Wiley & Sons, New York, USA.

Bateson, G. (1972). *Steps to an ecology of mind.* Chandler Publishing Co., San Francisco, USA.

Beer, S. (1979). *The heart of enterprise.* John Wiley & Sons, Chichester, UK.

Bullock, M. G. (1979). *Work process control guide.* SSDC-15. System Safety Development Center, EG&G, Idaho Falls, USA.

Dewey, J. (1938). *Logic: The theory of inquiry.* Holt, Rinehart and Winston, New York, USA.

Espejo, R., W. Schumann, M. Schwaninger and U. Bilello (1996). *Organizational transformation and learning - a cybernetic approach to management.* John Wiley & Sons, Chichester, UK.

Guldenmund, F. (2000). The nature of safety culture: A review of theory and research. *Safety Science,* **34,** 215-257

Hale, A. R., B. Wilpert and M. Freitag. (eds.) (1997). *After the event: From accident to organizational learning.* Pergamon, London, UK.

Kingston, J. (2001). *Organisational learning from incidents.* John Kingston Associates, UK

Koornneef, F. (2000). *Organised learning from small-scale incidents.* Delft University Press, Delft, The Netherlands.
www.library.tudelft.nl/dissertations/diss_html_2000/tpm_koornneef_20000926.html

Nertney, R. J. (1987). *Process operational readiness and operational readiness follow-on.* SSDC-39. System Safety Development Center, EG&G, Idaho Falls, USA.

Schaaf, T. W. v. d., D. A. Lucas and A. R. Hale (eds.) (1991). *Near-miss reporting as a safety tool.* Butterworth-Heineman, Oxford, UK.

Schein, E. H. (1992). *Organizational culture and leadership.* Jossey-Bass, San Francisco, USA.

Senge, P. M. (1990). *The fifth discipline: The art and practice of the learning organization.* Doubleday, New York, USA.

Westrum, R. (1991). Cultures with requisite imagination. In: *Verification and validation in complex man-machine systems* (Wise, J., P. Stager and J. Hopkins, eds.), Springer, New York, USA.

3

ORGANISATIONAL LEARNING AND KNOWLEDGE MANAGEMENT

Marleen Huysman [1]

INTRODUCTION

Knowledge management can be considered a management tool to support organisational learning. Linking knowledge management with organisational learning is helpful as it connects the various levels within an organisation: individual, group and organisation. When knowledge management is held up as being the management of organisational learning processes, then the link with the organisation as a whole and its organisational goals is made clearer. Organisational learning involves different types of knowledge-sharing. It is by sharing knowledge - in whatever form this might take - that organisations learn - irrespective of the way in which this takes place and the results obtained. In order to set organisational learning processes on the right track - in other words, to manage organisational learning processes in such a way that they contribute to the organisation's goals - it is of paramount importance both to support existing knowledge-sharing processes and to initiate new ones.

In this chapter, we will discuss this linkage between knowledge management and organisational learning in more theoretical detail. We begin this chapter by examining several theoretical premises upon which the concept of organisational learning is based. A theoretical framework is drawn up based on the social constructivist approach. This framework will serve

---

[1] Vrije Universiteit Amsterdam, The Netherlands

as an analytical model when the knowledge-sharing processes that are considered to form part of organisational learning are analysed.

THEORIES ON ORGANISATIONAL LEARNING

In this section we provide a more systematic and conceptual understanding of the process of 'organisational learning'. Without doubt, management terms such as 'learning company' and 'the learning organisation', etc. have gained good currency among academics and organisational practitioners. One plausible explanation for this attention is that learning organisations are generally seen as the solution to problems caused by hierarchical and bureaucratic organisations. The concept of *learning organisation* generally refers to a *specific type of organisation* that is organized – both culturally and structurally – so that innovation, flexibility, and improvement can be guaranteed. Literature on the learning organisation perceives learning as being something worth striving for. Consequently, the literature predominantly focuses on providing best-practices and models in order for consultants and managers to intervene. Its argument in short is as follows: within today's turbulent environments, only learning organisations are able to survive and thus gain a competitive advantage (e.g. Garvin, 1993; Marquardt, 1996; Pedler *et al.*, 1991; Senge, 1992)

Ideas on "the learning organisation" are different from ideas on the process of "organisational learning". The following premise is typical of the literature that treats organisational learning as a process: Organisational learning is a basic element in the evolution of organisations; every organisation learns, regardless of how it operates. Whether or not this learning will result in the organisation improving or renewing its organisational knowledge ('good learning') cannot be determined beforehand and therefore continues to be a subject for research.

Various authors have argued that there is a growing dichotomy between these two streams of research: the learning organisation stream and the organisational learning stream (Easterby-Smith *et al.*, 1999 Huysman, 2000a Tsang, 1997). These two streams represent two almost contrasting perspectives (see table 3.1).

Despite its popularity, the ideas concerning the learning organisation more often than not lack a solid theoretical, as well as empirical, foundation. This is a clear disadvantage as insights into the way organisations learn are a necessary precondition to construct prescriptive arguments on how organisations should learn. In other words, in order to create a learning organisation that is good at organisational learning, we first need to have a more conceptual understanding about the processes of organisational learning. This kind of organisational learning perspective on the learning organisation seems to be a fruitful combination.

Table 3.1: Differences between the learning organisation and organisational learning

	Organisational learning	Learning organisation
Outcome	Potential organisational change	Organisational improvement
Motive	Organisational evolution	Competitive advantage
Writings	Descriptive	Prescriptive
Objective of writings	Theory building	Intervention
Stimulus	Emergent	Planned
Targeting audience	Academic	Practice
Scientific background	Decision theory, Organisation studies	Organisational development, Strategic management

In the rest of this chapter we discuss ideas taken from the approach that sees organisational learning as a process. By discussing ways to support this process so that it contributes to organisational improvement, our general discussion and perspective on supporting knowledge-sharing supports the 'organisational learning perspective on the learning organisation', as mentioned above.

Among the first proponents of such an approach were Cyert and March (1963). In their 'Behavioral theory of the firm' they argued that organisations learn by adapting their objectives, attention and search routines to their experiences. More than a decade later, March and Olsen (1976) showed that, as a result of often irrational organisational behaviour, learning is full of hindrances and shortcomings. Two years later the frequently quoted book of Argyris and Schön (1978) was published. Like March and Olsen, these authors also argued that actual learning processes in organisations seldom result in positively valued changes. Organisations seem to have problems in thinking and acting outside existing theories-in-use. In the following years many review articles were published analysing various publications on organisational learning (e.g. Dodgson, 1993; Fiol and Lyles, 1985; Hedberg, 1981: Huber, 1991)

All these and other efforts notwithstanding, there is still a need for more scientific understanding on how to explicate actual organisational learning processes (Thatchenkery, 1996). The concept makes analysis difficult. For example, the traditional behaviouristic approach to learning seems to be problematic when applied to organisational learning. The stimulus-response sequence, traditional to the behaviouristic approach, is difficult to unravel as the combination of same stimulus, different response is rare in organisations (Weick, 1979). Organisations are too routine-based to follow this traditional learning sequence (Leavitt and

March, 1988)). Also, organisations do not provide the optimal (experimental) research site to unravel stimulus-response sequences. Moreover, researchers have problems *seeing* organisations and likewise learning of organisations. If organisations cannot be perceived, than it will be difficult to theorize about them, let alone about process of organisational learning (Yanow, 2000; Sandelands and Srivatsan, 1993). Many researchers also have difficulty differentiating between individual and organisational learning. Argyris and Schön (1978) for example talk about organisations while in fact they are dealing with learning individuals within organisations.

We use a broad definition of learning that emphasizes organisational knowledge construction: "Organisational learning is the process through which knowledge in an organisation is constructed or existing knowledge is reconstructed". The focus is on collective knowledge construction and is in line with more recent contributions to the organisational learning research stream (e.g. Brown and Duguid, 1991; Cook and Yanow, 1993; Elkjaer, 1999; Gherardi, 2000; Gherardi, Nicolini and Odella, 1998; Huysman, 2000 (a or b); Nicolini and Meznar, 1995; Pentland, 1995).

Those who perceive organisational learning as a process of (re)constructing organisational knowledge have all been inspired by the social constructivist approach to knowledge development (Berger and Luckman, 1966; Gergen, 1994; Schutz, 1971). According to the social constructivist approach, organisational learning is seen as an institutionalising process through which individual knowledge becomes organisational knowledge. Institutionalisation is the process whereby practices become sufficiently regular and continuous collective practices as to be described as institutions. The attention is on the process through which individual or local knowledge is transformed into collective knowledge as well as the process through which this socially constructed knowledge influences, and is part of, local knowledge. With organisational or collective knowledge reference is made to knowledge as in rules, procedures, strategies, activities, technologies, conditions, paradigms, frames of references etc. around which organisations are constructed and through which they operate (Leavitt and March, 1988). Important is that organisational knowledge is capable of surviving a considerable turnover in individual actors.

Next to constructing knowledge from within, knowledge can be gained by adapting to the environment. This learning from other organisations takes shape by reacting on feedback information from the environment or through assimilating knowledge from other organisations. Organisational learning happens when this external knowledge is institutionalised within the organisation. In the rest of this chapter we will not discriminate between internal and external learning but will mainly focus on internal learning as being a manifestation of organisational learning.

THE PROCESS OF INSTITUTIONALISING KNOWLEDGE

The essence of organisational learning is the (re)construction of organisational knowledge such as organisational norms, procedures, technologies, stories etc. Through knowledge-sharing, individual knowledge may become collective (organisational) knowledge while this accumulated knowledge will in turn influence subsequent action. In other words: organisational learning can be looked upon as a process that occurs as a result of the actions of the organisation's members, while these same actions are simultaneously influenced by collectively accepted knowledge as laid down in the norms and values, rules and procedures, and systems, that is by existing organisational knowledge. As a result of this duality between, on the one hand, the actions of individuals and, on the other hand, the deterministic or formative influences of existing organisational factors, organisational learning can be viewed as a process of institutionalisation (Berger and Luckman, 1966). Berger and Luckman (1966) describe the process of institutionalisation as consisting of three phases or 'moments': 'externalisation, objectification, and internalisation'. These three moments have proven relevant when analysing organisational learning processes (Pentland, 1995; Huysman, 2000a) and thus can help us to understand knowledge-sharing processes better. Externalising refers to the process through which personal knowledge is exchanged with others. Objectifying refers to the process through which society becomes an objective reality. During internalising, "the objectified social world is retrojected into consciousness in the course of socialization". As such, the authors point to a dialectical relationship between action and structure: "*the relationship between man, the producer, and the social world, his product, is and remains a dialectical one. That is, man (not, of course, in isolation but in his collectivities) and his social world interact with each other. The product acts back upon the producer*" (Berger and Luckman, 1966, p. 78).

Relating these processes of institutionalisation with organisational learning makes it possible to analyse organisational learning as consisting of these three consecutive moments:

- *externalising* individual knowledge in such a way that individually held knowledge becomes communicated. In other words: the exchange of knowledge for purpose of re-use or renewal;

- *objectifying* this knowledge into organisational knowledge so that shared knowledge is eventually taken for granted. In other words: the collective acceptance of knowledge
- *internalising* this organisational knowledge by members of the organisation. In other words: retrieving organisational knowledge.

Figure 3.1 presents this institutionalisation process through organisational learning as a 'knowledge cycle'.

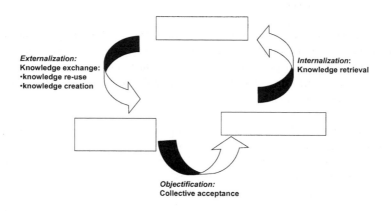

Externalization:
Knowledge exchange:
•**knowledge re-use**
•**knowledge creation**

Internalization:
Knowledge retrieval

Objectification:
Collective acceptance

Figure 3.1: Knowledge cycle

Internalisation: Acquiring organisational knowledge

When individuals acquire organisation's knowledge this is called 'internalisation'. It is through internalisation that individuals become members of the organisation and remain so. In fact, internalisation means the process through which one becomes an 'insider'. Acquiring knowledge takes place for example with the use of knowledge systems, training sessions, manuals etc. But it is also supported by the transfer of organisational knowledge that has not been recorded. Spender (1996) refers to this social knowledge that is of a tacit nature, as 'collective knowledge'. A powerful way to support the transfer of this knowledge is by telling stories or swaping anecdotes (Sims, 1999). Another way is by letting people work together. There is a growing group of authors that argue that learning should be considered as inextricable bound up with working (e.g. Brown and Duguid, 1991; Nicoloni and Meznar, 1995; Yanow, 2000; Gherardi, 2000). For example, Lave and Wenger (1991) introduced the concept of 'legitimate peripheral participation' as a method of learning by actively participating as opposed to learning outside the relevant task environment such as accumulating information from manuals.

Externalisation: Reuse or renewal

Knowledge exchange takes place between individuals. Individuals share their knowledge with other people in the organisation and this in turn begets shared knowledge. During the process of externalisation personal knowledge is transferred to others. Externalisation takes place in different ways; via formal channels such as meetings and project groups as well as through informal channels such as conversations in the corridors.

The externalisation of individual knowledge is facilitated when the knowledge is explicit by nature. Knowledge that can be expressed in language is only the tip of the iceberg (Polanyi, 1958). Knowledge can vary in the extent to which it can or cannot be passed on. Almost all knowledge has an explicit and an implicit or 'tacit' dimension. Explicit knowledge can be conveyed with the help of formal, systematic language. Implicit or 'tacit' knowledge is not formalized and extremely personal and therefore difficult to pass on. That personal knowledge is often highly implicit gives rise to obstructions in the externalisation process, which in turn can lead to substandard learning processes (Nonaka and Takeuchi, 1995).

Broadly speaking, there are two reasons why knowledge is externalised: knowledge exchange for the sake of reusing existing knowledge and knowledge exchange for the benefit of knowledge development. When knowledge is reused this involves knowledge 'flowing' from the knowledge carrier to the knowledge receiver. In the case of knowledge development it is not so much a case of one-way traffic but rather a reciprocal process of knowledge exchange.

Knowledge reuse involves adaptive learning process. However, if too much attention is given to this adaptation alone, conservatism or even inflexibility might result (Leonard, 1995). In the literature on how organisations learn this dichotomy between adaptation and innovation is often mentioned, albeit as a contrast between 'single loop learning' and 'double loop learning' (Argyris and Schön, 1978), 'adaptive learning and 'generative learning' (Senge, 1992), or 'exploitation' and 'exploration' (March, 1991). In every case, in the first learning method existing knowledge is adapted in such a way that it remains unaffected. The second learning method involves renewing this knowledge.

Objectification

Just because knowledge is exchanged does not mean that the shared knowledge has already been collectively accepted. In other words, shared knowledge only turns into organisational knowledge when it is accepted as such by the organisational members. This process of collective acceptance or objectification does not always takes place consciously and can be a long, drawn out process.

Collectively accepting local knowledge is the process in which the collective – often gradually – starts to accept existing shared knowledge as being part of the organisation. This process is not so much one of sharing knowledge but more one of sedimentation. Von Krogh, Ichijo and Nonaka (2000) refer to this process in the context of knowledge creation as 'globalising local knowledge'. For example, a group of technicians might have learned a new way of fixing a machine. This new operational knowledge remains local knowledge until it is accepted by the organisation, for example as expressed in organisational stories, in manuals and in the training of newcomers. This process of objectification usually takes much longer than is the case with the three knowledge-sharing processes discussed above (Berger and Luckman, 1966; Dixon, 2000; Douglas, 1987).

Most often, collective acceptance occurs when knowledge-sharing processes are ratified through the endorsement of dominant coalitions within an organisation (March and Simon, 1993). By addressing the role of dominant coalitions in supporting knowledge-sharing, we include the notion of power in the discussion on knowledge management. Power plays a crucial role during the objectifying process. Dominant coalitions are formed by, for example, management, a critical mass, reference groups, old-timers, or charismatic personalities. Dominant coalitions can have a negative impact on the result of learning processes. For example, management – as an important member of a dominant coalition - might be oblivious to what is actually going on within the organisation. By not accepting existing knowledge as important to the organisation as a whole, management hinders construction of organisational learning. As a result of which the learning process of the organisation is out of step with the learning process of individuals within the organisation (Brown and Duguid, 1991).

Knowledge-sharing only supports the process of organisational learning when the different practices are followed by collective acceptance or rather knowledge objectification. Collective acceptance as a process is, in other words, the link between individual learning and organisational learning.

As mentioned earlier, learning at the level of the organisation only takes place when the collective treats knowledge as being organisational knowledge, that is, when knowledge has become collectively accepted and used. Collectively accepting knowledge is of strategic importance. Organisations who want to exploit internal knowledge-sharing should pay particular attention to the collective acceptance of shared knowledge. This process is – or rather should be – an integral aspect of the other three knowledge-sharing processes. In chapter 13 of this book, the three types of knowledge sharing are illustrated with case study results.

SUMMARY

This chapter introduced some theoretical notion surrounding the concept of 'organisational learning'. We believe the concept offers a good framework that can be of help when analysing knowledge-sharing. When we focus attention on learning by the organisation, instead of just learning by the individual, knowledge-sharing is then also given meaning for the organisation as a whole. Depending on the learning process, knowledge-sharing takes on different forms. Knowledge acquisition occurs when individuals learn from organisational knowledge. Knowledge exchange takes place when individuals learn from each other. Knowledge development occurs when individuals learn *with* each other. In chapter 13 of this book the three forms of knowledge-sharing with the help of ICT are discussed in more detail.

REFERENCES

Argyris, C. and D. Schön. (1978). *Organizational learning: a theory of action-perspective.* Addison-Wesley, Reading, MA.

Berger, P. and T. Luckman (1966). *The social construction of knowledge.* Penguin books, London.

Brown, J. S. and P. Duguid. (1991). Organizational learning and communities of practice: towards a unified view of working, learning and innovation. *Organization Science*, **2** (1).

Cook, S. D. N. and D. Yanow. (1993). Culture and organizational learning. *Journal of Management Inquiry*, **2** (4).

Cyert, R. M and J. G. March (1963). *A behavioral theory of the firm*. Prentice Hall, Englewood Cliffs.

Dixon, N. M. (2000). *Common knowledge*. Harvard Business School Press, Cambridge MA.

Dodgson, M. (1993). Organizational learning: A review of some literatures. *Organization Studies*, **14** (3), 375-394.

Douglas, M. (1987). *How institutions think*. Routledge and Kegan Paul, London.

Easterby-Smith, M. L. Araujo and J. Burgoyne (eds.) (1999). *Organizational learning and the learning organization: Developments in theory and practice*. Sage, London.

Elkjaer, B. (1999). Organizational learning: A management tool or part of human interaction? In: *Organizational learning and the learning organization: Developments in theory and practice* (M. Easterby-Smith, L. Araujo. and J. Burgoyne, eds.). Sage, London.

Fiol, C. M. and M. Lyles (1985). Organizational learning. *Academy of Management Review*, **10** (4), 803-813

Garvin, D. A. (1993). Building a learning organization. *Harvard Business Review*, **71** (4), 78-91.

Gergen, K. J. (1994). *Realities and relationships, soundings in social construction*. Harvard University Press, Cambridge MA.

Gherardi, S. (2000). Practice-based theorizing on learning and knowing in organizations. *Organization*, **7** (2), 211-223.

Gherardi, S., D. Nicolini. and F. Odella (1998). Towards a social understanding of how people learn in organizations. *Management Learning*, **29** (3), 273-297.

Hedberg, B. (1981). How organizations learn and unlearn. In: *Handbook of organizational design, vol. 1. Adapting organizations to their environments* (P .C. Nystrom and W. H. Starbuck, eds.). Oxford University Press, New York.

Huysman, M. H. (2000a). Organizational learning or learning organizations., *European Journal of Work and Organizational Psychology*, **9**, 133-145

Huysman, M. H. (2000b). Rethinking organizational learning. *Accountancy Management and Information Technology*, **10**.

Huber, G. P. (1991). Organizational learning: the contributing processes and the literatures. *Organizational Science*, **2** (1), 88-115.

Lave, J. and E. Wenger (1991). *Situated learning: Legitimate peripheral participation*. Cambridge University Press, Cambridge.

Leavitt, B and J. G. March (1988). Organizational learning. *Annual Review Sociology*, **14**, 319-340.

Leonard, D. (1995). *Wellsprings of knowledge: building and sustaining the sources of innovation.* Harvard Business School Press, Boston.

March, J. G. (1991). Exploration and exploitation in organizational learning. *Organizational Science*, **2** (1), 71-87.

March, J. G. and H. A. Simon (1993). *Organizations*, 2nd edition. Blackwell Publishers, Cambridge MA.

March, J. G. and J. P. Olsen (1976). *Ambiguity and choice in organizations.* Universitets-forlaget, Bergen, Norway:.

Marquardt, M. J. (1996). *Building the learning organization.* McGraw-Hill, New York

Nicolini, D. and M. B. Meznar (1995). The social construction of organizational learning: Conceptual and practical issues in the field. *Human Relations, 48* (7), 727-746

Nonaka, I. and H. Takeuchi (1995). *The knowledge creating company.* Oxford University Press, New York.

Pedler, M., J. Burgoyne and T. Boydell (1991). *The learning company.* Book Company, McGraw Hill.

Pentland, B. T. (1995). Information systems and organizational learning: The social epistemology of organizational knowledge systems. *Accounting, Management and Information Technology*, **5**, 1-21.

Polanyi, M. (1958). *The tacit dimension.* New York: Anchor.

Sandelands, L. E. and V. Srivatsan (1993). The problem of experience in the study of organizations. *Organization Studies*, 1-22.

Schutz, A. (1971). *Collected papers*, vol 1 and 2. Nijhoff, The Hague.

Senge, P. (1992). *The fifth discipline, the art and practice of the learning organization.* Randon House London, London.

Sims, D. (1999). Organizational learning as the development of stories. In: *Organizational learning and the learning organization: Developments in theory and practice* (M. Easterby-Smith, L. Araujo. and J. Burgoyne, eds.). Sage, London.

Spender, J. C. (1996). Making knowledge the basis of a dynamic strategy of the firm. *Strategic Management Journal, ***17**, 45-62

Thatchenkery, T. J. (1996) Organizational learning, language games and knowledge creation. *Journal of Organizational Change Management. ***9** (1), 4-11.

Tsang, E. (1997). Organizational learning and the learning organization: A dichotomy between descriptive and prescriptive research. *Human Relations*, **50** (1), 73-90.

Von Krogh, G., Ichijo, K. and Nonaka, I. (2000). *Enabling knowledge creation.* University Press, Oxford.

Weick, K. E. (1979). *The social psychology of organizing.* Reading, MA: Addison-Wesley.

Yanow, D. (2000). Seeing organizational learning: a cultural view. *Organization,* **7** (2), 247-268.

4

KNOWLEDGE TRANSFER MECHANISM

Steven Dhondt [1]

INTRODUCTION

The current discussion on knowledge management is not a neutral discussion. It is mainly technology oriented, in this sense that only the technological possibilities are examined and less the social consequences. Also, little attention is directed at the limited use of implemented knowledge management systems (Damodaran *et al.*, 2000) or to the limited support knowledge management systems have to the goals companies have (Strikwerda, 2000; see also Huysman in this volume). This technological interest is spurred by the new possibilities in knowledge management systems. The reorientation of these systems from pure central databases to active collecting and connecting systems (e.g. by obliging connected workers to deliver inputs on lessons learned, performance etc.) is mainly responsible for this development. This interest on knowledge clouds the fact that knowledge is also power. The development of knowledge management systems is therefore not neutral for those affected by the system: the changes in knowledge generation is also a power shift in the company. This shift is also intended by management.

This chapter develops a perspective in which knowledge transfer mechanisms within companies are linked to power relationships. We start with distinguishing the core

---

[1] TNO Work & Employment, Hoofddorp, The Netherlands

characteristics of knowledge creation. Some of these ideas are in line with the elements presented by Koornneef (see chapter 2). From the core elements of knowledge creation, we can then look at the central knowledge transfer mechanisms and see how they can be related to power relations in companies. In chapter 12 of this book, four case studies are analysed using the perspective developed in this chapter.

KNOWLEDGE CREATION

Our point of departure is the question why management should be interested in controlling knowledge creation in their companies. The fact is that during most of the last century, the tayloristic and fordistic organisational forms gave management an absolute control on knowledge creation in companies. Knowledge creation was singled out into staff or research departments. But these tayloristic organisations have shown quite some weaknesses in the new competitive environments of the 1980s and 1990s. The clearest weakness was the small ability of these organisations to innovate and to adapt to new situations.

Staff control gave the possibility to standardise production to an extreme degree. The result of this standardisation tendency was that at the end, production was reduced to reacting to all those situations which weren't standard. The further development of the capitalist production model has lead to a situation wherein most productive tasks are reduced to 'reaction to non-standard situations'. Most operational and day-to-day tasks have been automated, so workers are left over with mainly exceptional disturbances. Reaction speed to these disturbances form the core of competitive firms. With this assessment, we touch the main problem with knowledge management. Figures 4.1 and 4.2 make this argument quite clear.

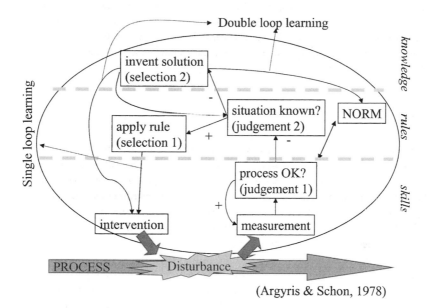

(Argyris & Schon, 1978)

Figure 4.1: Double loop learning explained with the control cycle (Argyris *et al.*, 1978).

Figure 4.1 shows how day-to-day problems in a work situation are solved. When workers are confronted with a disturbance (a deviation from 'normal processes'), they rely on rules to solve the situation. If no rules are available, workers try to develop new rules to solve these problems or adapt the norm. If only rules are developed, single loop learning is the case. Adaptation of norms is double loop learning. Trial and error play an important role in selecting the right solution. Figure 4.2 shows which sources of disruptions can be distinguished.

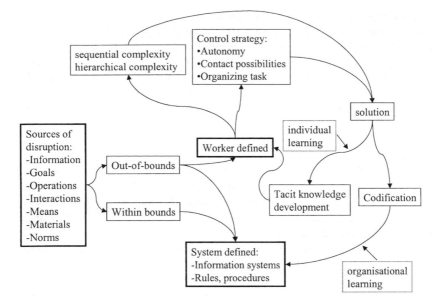

Figure 4.2: Procedures followed to control disturbances in production environments.

Figure 4.2 shows how workers solve those exceptions when they cannot rely on old rules. Disruptions which are 'known' are called 'within bounds'. For these disruptions, there are rules or there are solutions which have been included into the information systems. For disruptions which are 'out-of-bounds', new rules must be developed or old norms must be adapted. To develop new rules and/or change norms, workers must rely on the way their tasks are organised. A first necessity for single and double loop learning is that the tasks make it possible to evaluate the results of previous actions and that the worker has the possibility to select 'suited' methods to attack the problem. According to Hacker *et al.* (1983), if such tasks possess this quality, they can be called sequentially complete. A second necessity is that workers need to be accustomed to handling such disruptions. Jobs which are ruled by clear rules how to deal with disruptions can be called 'simple jobs'. The tasks in such jobs are monotonous and repetitive. If there are sufficient possibilities to deal with unforeseen situations in a non-standard way, then we can call the tasks and the job sufficiently complex. This complexity is called 'hierarchical complexity' (Hacker *et al.*, 1983). A third necessity is the degree in which workers can devise a control strategy (Vaas *et al.*, 1995). A suited control strategy means that the worker is able either to solve the problem himself (by means of changing the order of the tasks, working method, working place, sequence of tasks), either to have colleagues help the worker to solve the problem or either by having the boss/manager/team leader giving support to the worker. If the worker does not have these possibilities, he is stuck with unsolvable problems. Such problems can be seen as stress risks (Vaas *et al.*., 1995. When solving such disruptions and exceptions, the worker uses these possibilities. Depending on the quality of these possibilities, workers can select better or worse solutions.

These solutions add to the knowledge a worker has about a production system. Organisations can support workers in developing such solutions. If these solutions are not codified or translated into rules, then organisations are dependent on the worker to use his developed knowledge in future situations. We can call this individual learning. This knowledge is called 'tacit knowledge'. If solutions are codified, then we can say that the organisation has learned. But it is also important that the organisation can rely on workers to help other workers to use the newly acquired knowledge. Codification is not sufficient for an organisation to confront new situations. Knowledge transfer is also necessary.

The way disturbances are solved gives an insight into the quality of work (Vaas *et al.*, 1995; Dhondt and Vaas, 2000). Jobs and tasks can be qualified as of high quality if these jobs and tasks are organised in such a way that workers can solve these disturbances themselves. Jobs are of low quality, if there possess little control possibilities.

KNOWLEDGE TRANSFER MECHANISMS

As was said earlier, knowledge is power. With the previous information, it is clear that organisations can choose between the two different knowledge development methods, i.e. individual learning or codification, or create a mixed strategy. Companies need to be very careful in choosing one of these alternatives. Too much tacit knowledge reduces the possibilities for organisational learning, too much codification reduces the motivation of workers to help the company. Codification can help companies to control costs, but if worker motivation disappears, this strategy might not be cost effective. Nonaka and Takeushi (1994) have analysed four different strategies to cope with the said problem. Figure 4.3 shows these four strategies.

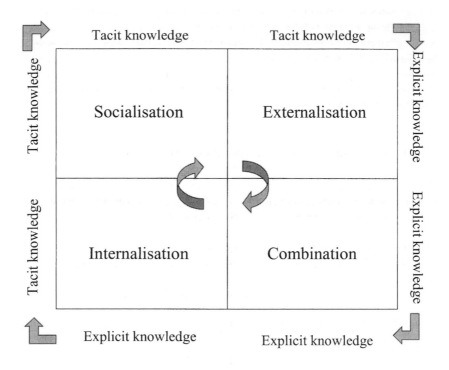

Figure 4.3: Four different strategies to cope with knowledge transfer.

The starting point of this model is the distinction between tacit and explicit knowledge. Explicit knowledge is what we have called codified knowledge. Its characteristics are: formally transferable, structured and systematised knowledge. Tacit knowledge is all non explicit knowledge such as experiences, insights, ideals, values and feelings. Both forms are transferred in different ways. Knowledge is created in an interaction between both types of knowledge. Four types of interaction can be distinguished:

- Socialization of knowledge: from 'tacit' to 'tacit'. Examples of this type of transfer are conversation, direct interaction, customer and supplier contacts;
- Externalisation of knowledge: from 'tacit' to 'explicit'. Examples of this type are formal presentations to peers, writing down of events, directives, procedures;

- Combination of knowledge: from 'explicit' to 'explicit'. Examples of this type are combining different written sources into a new one;
- Internalisation of knowledge: from 'explicit' to 'tacit'. Examples of this type are learning by doing, reading, courses, practice.

Companies have to be aware of these interactions. They should try to stimulate these four forms as much as possible.

ORGANISATIONAL TYPES AND KNOWLEDGE TRANSFER

This knowledge transfer can be related to organisational types. In table 4.1, we look into the dominant knowledge transfer mechanisms and organisational types. For organisational types, we use the Mintzberg typology (Mintzberg, 1983; Kraan *et al.*, 2002).

Table 4.1: Dominant knowledge transfer mechanism and organisational type.

Mintzberg (1983)	Sociotechnical typification	Dominant group	Dominant transfer mechanism
Simple structure	Centralised & concentrated control	Central management	Externalisation, Combination
Professional bureaucracy	Decentralisation & deconcentration	Production (operations)	Socialisation, Internalisation
Machine bureaucracy	Limited decentralised control ; high degree of concentration	(Technical) Staff	Externalisation, Combination
Adhocracy	Decentralisation & concentration	Staff	Socialisation, Externalisation
Divisional form	Limited decentralisation & deconcentration	Middle management	Externalisation, Combination

Table 4.1 shows how we hypothesize the correlation between organisation types as distinguished by Mintzberg (1983) with dominant knowledge transfer mechanisms. The simple structure is an organisational model in which all controlling power is centralised. Central management directs all decisions. Our hypothesis is that management will only be interested in externalisation as knowledge transfer mechanism. In the professional bureaucracy, the power resides in the production or service workers. An example is a hospital in which the physicians

have the controlling power. These professionals do not let themselves be controlled by central management. It is these professionals which create the new knowledge and who give on the knowledge to the colleagues. No explicit knowledge is created, rather the opposite, all explicit knowledge is internalised. These organisations depend on the will of the professionals to contribute to the goals of the organisation. Another model is the machine bureaucracy in which the technical staff have control of the organisation. It is the technical staff who decides how work and processes in the organisation need to be done. The technical staff is mainly interested in standardisation of work processes. This means that the dominant knowledge transfer mechanism is externalisation. New knowledge can be created by combination existing forms of knowledge. The prerequisite being that all knowledge is external. The adhocracy is much in line with the machine bureaucracy. The staff is however somewhat more 'friendly' towards the organisation in this sense that control is effectuated by trying to get into an agreement with the line organisation. The organisation is split into different centres which have autonomous decision power. In the divisional form, middle management is dominant. The divisions have some power which is delegated from the top. The knowledge transfer mechanism is oriented at externalisation, but also at combination of existing knowledge.

CONCLUSION

In this chapter, we developed a perspective in which knowledge transfer mechanisms within companies are linked to power relationships. We start with distinguishing the core characteristics of knowledge creation. Knowledge creation starts with reaction of workers to non-standard situations. Knowledge creation can lead to codified knowledge, but also to more tacit knowledge. For companies, it is necessary not only to rely on tacit knowledge, but to create systems of knowledge transfer. Four different transfer systems can be distinguished: socialization (from 'tacit' to 'tacit'), externalisation (from 'tacit' to 'explicit'), combination (from 'explicit' to 'explicit') and internalisation (from 'explicit' to 'tacit'). We hypothesize that there is a correlation between organisation types as distinguished by Mintzberg (1983) with dominant knowledge transfer mechanisms. Organisations with a simple structure centralise all controlling power so management will only be interested in externalisation as knowledge transfer mechanism. In the professional bureaucracies, professionals resist control by central management. No explicit knowledge is created, rather the opposite, all explicit knowledge is internalised by these professionals. In traditional machine bureaucracy, but also in adhocracy models, the technical staff control the organisation by means of standardisation of work processes. The dominant knowledge transfer mechanism in such organisations is externalisation. In the divisional form, middle management is dominant. The knowledge transfer mechanism is oriented at externalisation, but also at combination of existing knowledge. These hypotheses can help to clarify why a lot of knowledge management projects are not successful. Too little attention is directed at the fit between these management systems and organisational type. In chapter 12 of this book, four case studies are analysed using the perspective developed in this chapter.

REFERENCES

Argyris, C. and D. A. Schön (1978). *Organizational learning: A theory of action perspective.* Addison-Wesley, Reading, MA.

Damadoran, L. and W. Olphert (2000). Barriers and facilitators to the use of knowledge management systems. *Behaviour & Information Technology*, **19**, (6), 405-413.

Dhondt, S. and S. Vaas (2000). *WEBA Analysis. Manual.* TNO Work & Employment, Hoofddorp.

Hacker, W., A. Iwanowa and P. Richter (1983). *Tätigkeitsbewertungssystem.* Psycho-diagnostisches Zentrum, Berlin.

Kraan, K.O., E. Cox-Woudstra., J. Mossink and S. Dhondt (2002). De ERP-droom en de invloed op arbeid en organisatie; drie casestudies geanalyseerd op basis van Mintzberg. In: *Met het oog op de toekomst van de arbeid* (R. Batenburg, ed.), pp. 151-169. Elsevier Bedrijfsinformatie, Den Haag.

Mintzberg, H. (1983). *Structures in fives: designing effective organizations.* Prentice Hall, Englewood Cliffs.

Nonaka, I. and H. Takeuchi (1996). *De kenniscreërende onderneming.* Scriptum, Schiedam.

Strikwerda, J. (2000). De beperkte visie van de consultant. *Management Consultant*, **5**, 46-49.

Vaas, S. and S. Dhondt (eds) (1995). *De WEBA-analyse.* Alphen aan den Rijn: Samsom.

5

ORGANISATIONAL MEMORY

F. Lehner[1]

INTRODUCTION AND CONTEXT OF ORGANISATIONAL MEMORY

This chapter deals with the concept of organisational memory and with knowledge management systems (KMS). KMSs result from the application of advanced database and network technologies to support organizational learning and knowledge management approaches. Organizational Memory (OM, for a definition see section 5.2) and Knowledge Management (KM) are concepts well known from organization science and learning theory. These concepts are seen as instruments for systematic interventions into an organization's way of handling knowledge. Main goal of these interventions for an organization is to cope with changes in the business world (for a detailed discussion of the significance of OM see Stein, 1995):

- the increased complexity, dynamics, fragmentation and decentralization of knowledge or knowledge development,
- the increased complexity of organizational structures and the permanent need to change these structures,
- the increased amount of non-traditional information to be managed, e.g. (hyper-text) documents, links, multimedia documents, communication acts.

---

[1] University of Regensburg, Germany

Many approaches have been developed which claim to guide organizations to use their common or shared memory in a more efficient way (for an extensive survey of existing KM or OM approaches see Lehner 2000). Existing approaches focus on organizational issues and consider the OM as a resource, which has to be managed like capital or labour. With the advent of advanced database technologies (e.g. knowledge discovery and knowledge bases, data mining, distributed data base systems, multimedia and hypermedia data bases, intelligent agents as well as management and decision support systems) and net and communication technologies, especially the so-called "Intranet"- or "Web"-technologies, as well as specialized knowledge management systems, or e-learning environments exist to support organizational processes of generating, institutionalising, retrieving and disseminating knowledge.

For some time now under the rubric "organizational memory" (OM), both innovative and familiar concepts as well as highly promising systems have been proposed and tested (cf. e.g. Bannon and Kuuti, 1996; Buckingham Shum, 1997; Lehner, 2000; Morschheuser, 1997; Poveschi, 1998; Stein and Zwass, 1995; Wargitsch, 1997). The topic has taken on an intense sense of immediacy given the worldwide processes of restructuring in both economy and society. Relevant projects are already being carried out, especially in large corporations. Environmental dynamics and the pressure of competition that necessitate the development or the activation of new capabilities are paving the way for change. These adaptations occur automatically only in the rarest of cases, but presuppose (learning) processes. Important goals include elevating organizational efficiency and flexibility, overcoming growth limits and integrating organizations after mergers and acquisitions. The concentration on qualitative dimensions of organizational design gains increasing importance. One could look at this as an "expansion inwards", wherein new or previously unused potential and strengths should be developed.

ORIGINS AND THEORETICAL BACKGROUND OF ORGANISATIONAL MEMORY

In general the term memory denotes a system capable of storing things perceived, experienced or lived beyond the duration of actual occurrence, and then retrieving them at a later point in time. Learning is not possible without memory. Accordingly, organizational memory is repeatedly proposed as a prerequisite for organized leaning in this context. Thereby, however, the term "organizational memory" should in no way be considered analogous to a "brain" to which organizations have access. The term is simply meant to imply that the organization's employees, written records, or data "contain" knowledge that is readily accessible (cf.

Oberschulte 1996, 53). Several functions of "memory" are inherently present in every organization without software-technical support of any kind (e.g. in the form of search and recall processes as carried out by telephone surveys or brainstorming during meetings).

Discussion of the existence of organizational memory has a longer tradition than one might at first suspect. Traces can be found as early as the end of the 19th century. However, at the beginning of the 20th century interest was lost. The topic was almost completely forgotten, despite psychology's close relationship to organizational memory and the spectacular theories being developed in the fields of cognition, memory, information processing and artificial memory (cf. Wegner 1987, 185).

Just as there are terms for the individual level (e.g. individual psychology, the psychology of learning), there are various terms concerning memory and definitions of knowledge for the collective and organizational levels. Intensive scientific investigation of this topic began at the end of the 70s and beginning of the 80s and was mostly carried out in the USA. A very popular article in this field was written by Duncan and Weiss (1979, p86-87). Hedberg develops this knowledge-based approach further. In 1981 he is the first to introduce the term "organizational memory" (Hedberg, 1981): Organizational memory establishes the cognitive structures of information processing, the theory of action, for the entire organization. Parallel to this development, Kirsch coins the term "organizational knowledge base" (organisatorische Wissensbasis) for the German-speaking realm (in unpublished papers in 1974 according to Güldenberg and Eschenbach, 1996). Building on this, Pautzke in 1989 identified several layers of the organizational knowledge base, which provided an important starting point for further research and conceptualisations. Hence organizational memory was really not a terminus technicus, but rather part of a tradition of organizational development, organizational learning, and knowledge management.

Human memory is often used as model and metaphor for the organizational memory. Comparing characteristics of individual and organizational memories the following differences seem to be important:

Metaphor	Term	Functions and Examples
Machine	Storage	- Erasing the contents of memory requires external actuation
		- Data can be called up, duplicated, or recovered
		- The structure and form for storage is established; the behaviour of the system determined

		-	Growth of the knowledge bases occurs primarily quantitatively, i.e. as an increase in the amount of data
Organism	Memory	-	Erasing occurs as "forgetting" and is an automatic process
		-	Knowledge is reproduced or reconstructed as needed (i.e. when it is used)
		-	The structure and form for storage is not established; the behaviour of the system not determined
		-	The growth of the knowledge base does not happen via accumulation, but rather through reorganization or structural transformations in the knowledge base respectively.

Various management approaches and scientific disciplines played and continue to play a role in the development of the theory of organizational memory, some of which enjoy a long and respected tradition of their own.

- Organizational learning (OL) and learning organization (LO)
- Organizational intelligence (OI), competitive intelligence (CI)
- Knowledge management (KM) as well as the concept of organization as knowledge and/or information processing system
- Organizational change, management of change, innovation management
- Organizational culture
- Theory of the evolution of organizations

Naturally, organizational memory is not the only component that plays a part in these disciplines and approaches, the aims and purposes of which often differ markedly from one another. Nevertheless, organizational memory has a common and meta-status for all these concepts, but especially for organizational learning and knowledge management. It is essential for the understanding of organizational memory systems, and it represents simultaneously an important bridge and common denominator for the other disciplines. These concepts are discussed in greater detail in the literature (see Lehner, 2000).

Most attempts to define or explain basic properties of organisational memory prefer either the idea of organisational memory as a concept or organisational memory as a construct. These two dimensions' basic properties are each characterized by one of their proponents. Concerning the concept view: "OM is a concept that an observer invokes to explain part of a system or behaviour that is not easily observed" (Krippendorg 1975, quoted from Rao and

Goldman-Segall 1995, 333-334). Concerning the construct view: Organizational memory "... is the know-how of a business recorded in documents (reports, ideas, concepts, etc.)" (Morschheuser 1997, 19). Regardless of the position one assumes, organizational memory has to do with either something abstract (theory, explanatory model, thought schemata, concept) or something concrete (e.g. documents, data bases, knowledge base, repository).

The fact that there is still no clear or unified use of terms is a sign of the liveliness and novelty of the research topic. As the discussions have shown up to this point, the meaning of the terms is also not always identical because they originate in part from different disciplines and therefore also have different aims as far as knowledge is concerned. Figure 5.1 summarizes the current status of terminology usage as they are related to each other.

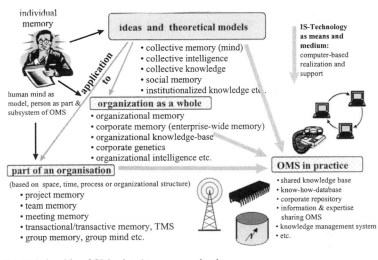

Figure 5.1: Relationship of OM-related terms to each other

FROM ORGANISATIONAL MEMORY TO ORGANISATIONAL MEMORY SYSTEMS – A PERSPECTIVE VIEW

Organisational memory systems are closely connected to the above presented ideas of organisational memory which attribute a particularly central role to organizational learning and knowledge management. However, the discussion on the different areas of application has not progressed much in the relevant literature. Due to the situation described above and the dynamic development of the application areas, it is not surprising, that there are hardly suitable definitions for this new type of information system yet. One of the few attempts at an explanation was made by Stein and Zwass (1995, p95), who define an OM Information System as "a system that functions to provide a means by which knowledge from the past is brought to bear on present activities, thus resulting in increased levels of effectiveness for the organization". This definition clearly aims at the contribution of such systems to the increase in organizational effectiveness.

Organisational memory systems and knowledge management systems[2] can be conceived as a special class of information system supporting the idea of organisational memory. They are particularly important for innovation and the retention of organizational flexibility. The ways and means in which information technology is actually used for particular corporate goals depend mostly on which concept of "organization" the business has as a model. The systems introduced into a corporation should correspond to the prevailing fundamental views of the time (technology-culture fit). A business that evolves according to the plan of a self-organising system (e.g. a community of practice) requires other software-technological solutions than a business that is run in a stable environment according to bureaucratic rules (e.g. a governmental office). Differences will also occur depending on how the information systems are used, even if they are the same system.

---

[2] Knowledge management systems in our view are a subset of organizational memory systems which have a tendency to focus on the more static documentation (retention, maintenance, search and retrieval) and distribution parts of organizational memory systems. This holds true for most of the systems offered on the market which lack functionality to support the dynamics of an organizational memory, that is organizational learning.

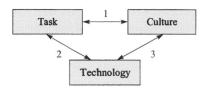

Figure 5.2: Corporate efficiency (is it efficiency or effectiveness?) through fit between tasks, culture and technology

By using different organizational measures, most modern managers attempt to create a climate in which learning in and about organizations is encouraged. Figure 5.2 shows the relevant components (Goodhue and Thompson, 1995). Most business management applications concentrate for the most part on the relationship represented by arrow 1, i.e. they attempt to create a fit between tasks or, in other words, between labour organization and the (organizational) culture. "Traditional" information management is mainly concerned with trying to bring tasks and the technology that has been introduced into line with each other (arrow 2). With the help of organizational memory systems, the attempt is made to additionally improve the relationship between the technology introduced and organization's culture (arrow 3) in order to achieve organizational efficiency. The most important concept for developing this new type of systems is organizational memory, which shall be explained more in detail.

The development of KMS or more general of OMS (organizational memory systems) is substantially more complex than the development of conventional information systems. In order to achieve the goals associated with OMS/KMS it is necessary to develop an understanding of the effected frameworks and management concepts that are already available. In addition, a more comprehensive understanding of the use of conventional database technology has to be developed and integrated within the broader concept of "organizational knowledge". One of the real challenges will be resolving the contradiction between construction paradigm (corresponding to technological systems) and the evolution paradigm (corresponding to the characteristics of social systems).

Even though there is considerable confusion about what exactly an organizational memory system is, both, researchers and practitioners agree on the importance and usefulness of the approaches to overcome the shortcomings of current practices of business engineering with respect to organizational effectiveness (see Roehl, 1999, p13; Stuart, 1996, p1). In order to solve the dilemma of a precise description, a perspective-based approach is introduced (see

Lehner, 2000). This allows different views to be regarded as explanations of equal importance without playing them off against one another. At the same time, it allows for the heterogeneous and even somewhat contradictory features associated with organizational memory systems to be systematically described. Until now this problem has not received sufficient attention in the literature. On the basis of the current level of knowledge, the following perspectives are proposed (see Lehner, 2000):

- OMS as a new kind of application systems
- OMS as a concept
- OMS in a functional view
- OMS as an attribute of information systems
- OMS in a behavioural view
- OMS in a technological view

The individual perspectives are not at all mutually exclusive and are briefly explained and summarized in the definitions below.

Perspective 1: OMS as a new kind of application systems

This perspective denotes OMS as new and highly specialized application systems prolonging the list of application systems including financial information systems, production planning systems, supply chain management, executive information systems and many others. An organisational memory system therefore will be a "real" information system realizing or supporting the organizational memory (e.g., a distributed database system connected through an Intranet or a workflow system with groupware functions). According to this understanding the following definition is suggested:

Definition 1: An organizational memory system (OMS) is a system that either creates parts of the organizational knowledge base using information and communication technology (class 1) and/or a system that creates or supports the tasks, functions, and processes associated with the use of organizational memory (class 2). The concept of organizational memory must be taken into consideration explicitly or implicitly in the objectives for the use of OMS as well as in the system architecture.

One difficulty with this perspective certainly occurs in delineating or defining the presupposed organizational knowledge base. At first glance, the definition also has the disadvantage that no conscious distinction is made to traditional information and database systems. On the basis of class 1, such traditional database systems would at least be categorized as OMS. Yet, including these systems in the definition is intentional and considered significant, since database systems make up an integral component of organizational memory. However, the difference and the added value compared to the traditional use of technology make organizational memory systems a topic of study in their own right.

Perspective 2: OMS as a concept

Objects have noticeable, concrete features that allow them to be described. Concepts have no such features. They are abstractions that are created or developed for specific purposes. Frequently, they serve to describe or analyse complex phenomena systematically. A well-known example of such a concept is human intelligence, or rather the intelligence quotient. With the help of the intelligence concept, particular observations of human abilities can be classified and scientific activities can be coordinated. The concept, however, is only significant in connection with these facts. Outside of this context, it does not exist (see Müller-Merbach 1996, 354). In a similar fashion, organizational memory systems can be understood as a concept. With the help of this concept, the analysis or selective operation of particular parts of an organization (e.g., structures or processes) are thought to be supported.

> *Definition 2:* An organizational memory system (OMS) is a concept that allows particular phenomena and capabilities in organizations to be described and explained. The latter are particularly linked to learning ability, intelligence, knowledge management, etc. The concept can be used to evaluate and improve the performance of these capabilities. The technical realization of sub-functions is included in this abstract concept.

Perspective 3: OMS in a functional view

Another way to understand organizational memory systems is to take as a starting point the functions that such systems perform (or should perform). Defining a system functionally is possible by referring to existing systems or to knowledge-management architectures (examples can be found in Glance *et al.*, 1998; Borghoff and Pareschi, 1998; Eulgem, 1998). All the suggestions concerning the basic functions of knowledge management found in professional publications can also serve as a starting point. On the basis of such an understanding, an OMS can be defined as follows:

Definition 3: An organizational memory system (OMS) is a computer-based system that, with the help of software, supports at least the following basic functions: the generation and acquisition, storage, search for and utilization of knowledge, as well as its distribution and updating.

The main difference to perspective 1 is the focus on functionality in perspective 3 while perspective 1 focuses on the retention facility (storage and organisational memory).

Perspective 4: OMS as an attribute of information systems

It should already be clear from what has been said above that OMS do not necessarily have to be systems that support a clearly defined purpose or task. The designation "OMS" may also be regarded as an attribute or characteristic that, next to other attributes (e.g., decision-making and group support), belongs to a system. The attribute itself, as well as its level of importance, may vary. It is useful, among other things, when it is important to determine the contribution that existing information systems make to organizational memory. In other words, it underlines the overall significance of this contribution.

Definition 4: An information system is called an organizational memory system (OMS) when it supports the search for, automatic storage and retrieval of a portion of information as well as explicit knowledge required in the process of determining a company's performance. The OMS qualification does not prevent other features and designations from being used to define the system more precisely.

Perspective 5: OMS in a behavioural view

With the behaviour-oriented perspective, an especially important aspect is stressed, namely the influence that information technology has on behavioural patterns. Here, the behaviour of both the individual as well as that of entire organizations (collective behaviour) is meant. The link between technology and behaviour, or rather behavioural changes, has been a subject of investigation for a long time, but consensus has yet to be reached on the direction of that influence. In connection with OMS, the situation is looked at more closely, since the principal concern is not changes brought about by technology, but rather attainable effects on the behavioural level. The following definition is proposed:

> *Definition 5:* First of all, organizational memory is defined as the totality of all components, data, documents, events, information, functions, mental concepts, and other entities in an organization that influence the particular behaviour or the behavioural disposition of the organization's members. An OMS is a computer-based system used to create or support those components and functions of the organisational memory directly influencing the behaviour and cognitive abilities of single persons, groups or entire entities of the organization. Therefore, one can also speak of an electronic environment that provides stimuli relevant to behaviour.

Perspective 6: OMS in a technological view

The technological understanding of OMS is probably the easiest way to understand such systems. Implied here is that certain technologies exist that are either developed or used for these kinds of systems. Such a perspective corresponds to an extent to the technological concept of organizational memory. The following definition is proposed:

> *Definition 6:* An OMS is a system developed using dedicated technologies or tools. Among these are, in particular, document-management systems, database systems, data warehouse, OMS tools such as Fulcrum, Answer Garden or Knowledge Garden, and platforms such as Lotus Notes, as well as combinations of these.

In accordance with these views or explanations, a company may have (or does have) several parallel OMS. These systems can operate independently of each another or be connected through a network (e.g., by means of technical interfaces, overlapping at the user level, or common areas of knowledge and application).

Summing up, the perspective-based view is intended to provide orientation in a dynamic research field by means of an instrument that can be used to position concrete research projects or questions. Organizational memory systems (OMS) are generally characterized by the fact, that a whole bundle of tools and systems is used and not an isolated single tool or system. A particular obstacle regarding the categorization occurs due to the close connection to the concepts of information and knowledge. These terms are usually understood in a very inhomogeneous and wide way, and a homogenisation is neither to be expected in the near future nor realistic. So it is not surprising that a very wide scope of realization forms is seen.

To come back to the examples at the begin of this chapter the spectrum reaches from a very restricted understanding in the context of specialized systems for the support of specific, relatively clearly defined tasks (e.g. the admission procedure of new medicines), via the company-wide use (e.g. knowledgeLINK), to very general information systems which at first have no specific connection to the corporate performance (e.g. Answer Garden, see Ackerman 1990).

REFERENCES

Ackerman, M. S. (1994). Definitional and contextual issues in organizational and group memories. In: *Proceedings of the 27th Hawaii International Conference of System Sciences (HICSS)*, Organizational Memory Minitrack.

Bannon L. and K. Kuuti (1996). Shifting perspectives on organizational memory: From storage to active remembering. In: *Proceedings of HICSS'96, 29th Hawaii International Conference on System Sciences, Hawaii, IEEE.*

Borghoff, U. M. and R. Pareschi (ed.) (1998). Information technology for knowledge management, Berlin.

Buckingham Shum, S. (1997). Negotiating the construction and reconstruction of organisational memories, *Journal of Universal Computer Science*, **3** (8), 899-928.

Duncan, R. B., and A. Weiss (1979). Organizational learning: Implications for organizational design. In: *Research in Organizational Behaviour* (B. Staw, ed.), pp. 75-123, *Greenwich, Conn.*

Eulgem, S. (1998). Die Nutzung des unternehmensinternen Wissens. In: *Perspektive der Wirtschaftsinformatik*, Frankfurt.

Glance, N. *et al.* (1998). Knowledge pump: Supporting the flow and use of knowledge. In: *Information Technology for Knowledge Management* (U. M. Borghoff and R. Pareschi, eds.), pp. 35-55, Berlin.

Goodhue, D. L., and R. L. Thompson (1995). Task-technology fit and individual performance. *MIS Quarterly*, **19**, 213-236.

Güldenberg, S. and R. Eschenbach (1996). Organisatorisches Wissen und Lernen – erste Ergebnisse einer qualitativ-empirischen Erhebung. *zfo*, **1**, 4-9.

Hedberg, B. (1981). How organizations learn and unlearn. In: *Handbook of Organizational Design* (P.C. Nystrom and W. H. Starbuck, eds.), pp.3-27, New York.

Lehner, F. (2000). *Organisational Memory. Konzepte und Systeme für das organisatorische Lernen und das Wissensmanagement*, Munich, Vienna.

Morschheuser, S. (1997): *Integriertes Dokumenten- und Workflow-Management*. Wiesbaden.

Müller-Merbach, H. (1996). Die „Intelligenz" der Unternehmung: Betriebliches Gestalten und Lenken aus einer neuen Sicht. In: *Umbruch und Wandel. Herausforderungen zur Jahrhundertwende* (C. P Claussen *et al.*, eds.), pp. 353-366, München.

Oberschulte, H. (1996). Organisatorische Intelligenz - Ein Vorschlag zur Konzeptdifferenzierung. In: *Managementforschung*, vol. 6: Wissensmanagement, (G. Schreyögg and P. Conrad, eds.), pp. 41-81, Berlin.

Pautzke, G. (1989). *Die Evolution der organisatorischen Wissensbasis. Bausteine zu einer Theorie des organisatorischen Lernens.* München.

Poveschi, R. (1998). *Information technology for knowledge management.* Berlin.

Rao, V. S. and R. Goldmann-Segall. (1995). Capturing Stories in Organizational Memory. In: *Proceedings of the 28th Hawaii International Conference on System Sciences '95, IEEE Press, Los Alamitos 1995*, 333-341

Roehl, H. (1999). Kritik des organisationalen Wissensmanagements. In: *Organisationslernen durch Wissensmanagement*, (Projektgruppe wissenschaftliche Beratung, eds.), pp. 13-37 Frankfurt/Main.

Stein, E. W. (1995). Organizational memory: Review of concepts and recommendations for management. *International Journal of Information Management,*. **15**, (1), 17-32.

Stein, E. and V. Zwass (1995). Actualizing organizational memory with information systems. *Information Systems Research*, **6**, (2), 85-117.

Stuart, A. (1996). 5 uneasy pieces – Part 2: Knowledge management. *CIO Magazine*, June 1st, 1996, http://www.cio.com/archive/060196_uneasy_1.html/

Waesche, N. and L. Guntrum. (1996). Realisation einer Intranet Kultur - am Beispiel des Hoechst Wide Web. *Industrie Management*, **12**, 6/1996, 39-42

Wargitsch, Ch. (1997). *Ein Organizational-Memory-basierter Ansatz für ein lernendes Workflow-Management-System.* Forschungsbericht FR-1997-004, Bayerisches Forschungszentrum für Wissensbasierte Systeme, Erlangen/München/Passau, 1997

Wegner, D. M. (1987). Transactive memory: A contemporary analysis of the group mind. In: *Theories of Group Behaviour* (B. Mullen and G. R. Goethals., eds.), pp.185-208, Springer Verlag, New York.

PART 2

LEARNING FROM ERRORS

6

IMPROVING KNOWLEDGE SHARING AND LEARNING IN AN ORGANISATION OF SAFETY, HEALTH AND ENVIRONMENTAL PROJECT ENGINEERS

Urban Kjellén[1]

INTRODUCTION

The object of the study presented in this chapter is a group of some twenty engineers employed by a company and involved in the management of safety, health and environment (SHE) during the planning and execution of the company's investment projects. The engineers' area of responsibility covered the follow-up of SHE in the design of oil and gas installations and industrial plants. SHE in this case included the implementation of an adequate design to prevent major and occupational accidents, to ensure a satisfactory working environment and to reduce environmental pollution. It also covered safety and environment during the construction of contract objects. The engineers' activities were important to the company's stakeholders for several reasons (Kjellén, 1998; Kjellén, 2000). Design has considerable effects on the SHE performance of the installations and plants during production. Timing of the implementation of SHE measures in design is also important from a cost point of view. Design measures to improve SHE to an acceptable level are usually much more costly if they are implemented at a late project phase or during operations than if they are implemented when design is made in the first place. A professional and smooth handling of contacts with authorities reduces risks of

---

[1] Norsk Hydro ASA, Oslo and NTNU, Trondheim, Norway

unplanned cost-increases or schedule delays (or even termination of the project). Adequate management of SHE in construction will reduce the risks of accidents that may result in injury to personnel, damage to the environment or contract object or schedule delay.

Nature of the Problem

The engineers belonged to a 'basis organisation', i.e. an Environment and Safety department within the Technology Centre of a company (see figure 6.1). A large majority of the engineers were located full-time in different project groups in such positions as SHE mangers, safety discipline lead engineers, and safety, working environment and environmental care engineers. A small group of engineers was physically located at the home office of the Technology Centre and conducted discipline studies such as risk analyses and environmental impact assessments.

A project usually lasted for several years. The engineers moved from one project to the next, and it took a relatively long time for an individual engineer to get experience from all project phases. The projects involved different types of plants and installations, and new technology and new management principles were implemented all the time. The individual engineers often met situations that they had not experienced before. This problem had to be addressed through provisions for adequate and systematic experience exchange between department members.

Early in 1999, a quality improvement project was commissioned to carry out an evaluation of the projects' SHE management function. It concluded that there was a certain degree of arbitrariness in the way SHE management was carried out in the different projects. The group pointed at a number of problems of relevance here. There were unclear expectations as to the SHE manager's role and no clear role model existed. Routines and work methods in project work had not been adequately defined. This had to do with an insufficient experience transfer between projects and incomplete standards and best practices for SHE management. The SHE managers lacked sufficient incentives to prioritise work for the 'basis organisation' (i.e. the Technology Centre) in developing such standards and best practices. It followed that individual SHE managers were allowed to shape their own role. There were variations between the individual SHE managers as to how they prioritised the different areas of SHE and at what level of detail they operated. The Technology Centre's follow-up of the project work through design reviews and audits was insufficient.

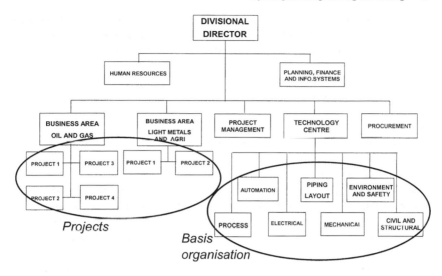

Figure 6.1: Simplified organisational diagram showing the basis organisation and the projects.

Although the quality improvement project addressed the SHE management position in particular, there were reasons to believe the identified problems applied to other project SHE positions as well. Already before the report was issued, a number of measures had been initiated to improve knowledge sharing and learning in the department. The report substantiated the needs for such initiatives.

This Study

The study addresses measures taken during 1998 and 2000 to improve knowledge sharing and learning within the Environment and Safety department of the Technology Centre. The measures aimed at stimulating the individual department members to improve the quality of their own work. They also aimed at capturing the department members' knowledge and to transform it into best practices to be useful to the organisation as a whole. More specifically, the measures aimed at supporting a uniform treatment of SHE issues in the projects, an adequate coverage of the relevant areas of SHE (i.e. process safety, working environment and environmental care and construction site safety) and cost consciousness. They should also promote a trustful relationship with the authorities and look after customer (i.e. the operations organisations) satisfaction.

The following measures are evaluated here:
A.   The department's professional networks
B.   A database on Best Practices
C.   Technical specifications and other steering documents
D.   Peer auditing

The chapter describes how the different measures developed in the course of a three years period. It also presents results of an evaluation in the form of a questionnaire study of the department employees. The aim of the chapter is to evaluate how an organisation of project SHE engineers can learn from each other's successes and failures. It addresses the question of the extent to which the different measures have been put into use. It also assesses the effects of the measures on the quality of the SHE work, competence development and feeling of solidarity among the members of the department. Finally, the chapter addresses the question of whether the context has been an adequate enabler of knowledge sharing and learning.

METHOD

Study Object

In the year 2000 the Environment and Safety department consisted of some 20 own employees and 5 external consultants. Slightly less than half of the employees had been hired during the last four years and a group of about the same size had a seniority of 10 years or more.

About three-quarters of the personnel were allocated full-time to investment projects primarily in the offshore oil and gas sector (see figure 6.1). These people were located in different parts of Norway and also in other countries in Europe and the Middle East. Their duties shifted when the projects progressed from the study to the execution phases (Kjellén, 2000). The engineers participated in design development during the study phases. They became more detached during project execution, and their main duty was to follow-up engineering, procurement and construction contractors to ensure that the authorities' and the company's requirements to SHE were met.

The remaining personnel had their daily workplace at the home office in Oslo and performed work on work orders for different projects and for other divisions. Personnel at the home office and projects used telephone and e-mail as their primary means of communication.

The department was lead by a department manager, reporting to the division's Technology Centre manager. So-called discipline responsible personnel had also been appointed for the core discipline areas of the department, i.e. technical safety, working environment, environmental care and HVAC (heat, ventilation and air-condition). They were members of the department's management team and were responsible for the discipline networks and for the development of methods and tools (i.e. Best Practice). The discipline responsible employees also played an important role as coaches of junior department members.

New employees usually started as a junior specialist within one of the department's core areas. In the course of their career, the engineers were offered opportunities to widen their area of competence within SHE management and within other areas of SHE. Still, the employees' identity by and large remained with the original core area. Personnel with a technical safety background was dominating and comprised about two thirds of the personnel.

Data Sources

This study is based on the following data sources:
- Working environment surveys in 1998 and 2000. The surveys were conducted by the division's Human Resource department and involved the whole division. This made it possible to compare the developments in the SHE department with other parts of the organisation.
- A questionnaire study and individual interviews of the department members in May 2000. These were carried out by an external consultant in organisational development and aimed at evaluating the preconditions for team building and co-operation in the department.
- A questionnaire survey in February 2001 of the department members' attitudes towards different measures to improve knowledge sharing and learning. 21 of the department's 22 employees participated in the study (95%). This was followed by a presentation of the survey results at a department seminar and group work.
- Minutes and other documentation from department seminars and meetings during 1998 – 2000.
- The author's own experiences as department manager in the period 1998-2000.

POINTS FOR INTERVENTION

Chronological events (mainly 1998 - 2000)

The department was originally established in the early 1980s to organize SHE personnel working in projects. Early in 1998, there was a change in management of the department. The new manager introduced the management team concept and the functions of discipline responsible. Two other early initiatives involved re-vitalization of the work on the department's Best Practice database and of the department networks. Hiring of new staff started early in 1998 and continued until early 1999. The reason was that the company planned large increases in its investment portfolio. In all, five new employees were hired.

Starting in 1998, the steering documents of the department (technical specifications, procedures and work instructions) were systematically updated and supplemented to reflect new work processes and organisational changes.

In February 1999, large cut backs in the company's investment portfolio were announced. In June the same year, it was announced that the company had acquired parts of another Norwegian Oil Company and that the project organisations of the two companies had to merge. Reorganisation and downsizing started in August 1999 and the new organisation took effect from January 1st 2000. The new organisation did not differ significantly from the earlier one, and the SHE department remained at about the same size. About a third of the employees left the department and new personnel from the acquired Oil Company joined the department.

Early in 1999, a quality improvement project was established to evaluate the SHE management function in the projects. The group consisted of three SHE managers from the department, one project manager and a representative from a client (i.e. an operations manager). The group presented their findings at a meeting with the department's SHE managers in June 1999 and in their final report in August 1999. Although findings and recommendations were followed up at the department level, the main message of the report was drowned by the noise around the reorganisation and down sizing.

In September 1999, all department members were offered the opportunity to participate in a three days' training in SHE auditing.

A team-building seminar was arranged in June 2000. Before the seminar, a consultant carried out interviews and a survey study. A number of actions were launched as a result of the seminar to improve co-operation and competence development. The formal discipline networks were reinforced, partly with new leadership. Two quality-improvement projects were initiated, one on the department's vision and the other on competence development among the employees.

There was a change in management of the department, effective from January 1st, 2001. The questionnaire to evaluate the different means to improve knowledge sharing and learning presented below was distributed in February 2001 and results were presented at a department seminar the following month.

Professional Networks

One of the conclusions from the quality improvement project in 1999 was that experience exchange between the projects was unsystematic. The professional networks had an important role in this context. We will here focus on the formal networks for department members working either in the different projects or at the home office. Before 1998, there were two such networks: one for the SHE managers in the different projects and the other for the environmental care engineers. Neither the technical safety discipline personnel, nor the working environment experts did have regular meetings.

Three problems were identified in relation to this network structure:
- There were disciplines and issues that were not covered by the network structure;
- Some of the employees and consultant did not participate in any of the professional network meetings; and
- The SHE managers' network did not give full attention to all areas of SHE. The meetings were by and large oriented towards technical safety issues, and other areas of SHE and the management aspect were de-emphasized.

Stepwise, a revised network structure took form. Work with the Best Practice database was revitalized in 1998. Each discipline responsible gathered a group of experts within the department for this work. They had seminars and meetings. Whereas these groups in practice formed discipline networks, some of the department members and the consultants did not participate in any such group.

An important milestone was reached after the summer of 2000, when all department employees and consultants were assigned to at least one network. The aims of the networks were made more precise. They were:

- To contribute to a qualitatively fully acceptable and reproducible treatment of SHE and HVAC issues in studies and projects;
- To provide an arena for experience exchange and development of the Best Practice database;
- To contribute to learning at the individual and organisational levels and to the development of new ideas;
- To contribute to the individual department members' personal networks and feeling of belonging with the department.

The networks themselves appointed a chairman mainly amongst those working in the projects. The SHE management network used a fixed agenda as a checklist to ensure that all relevant issues were covered in the meetings. Initiatives were also taken to invite SHE experts from of the client to the network meetings.

At the end of 2000, there were five networks of varying size, ranging from three to seven members.

Database on Best Practice

The quality improvement project concluded in 1999 that the standards and best practices for project SHE management were incomplete. The database on Best Practice was originally named 'the department tool-kit' and could be traced back to 1996, when a work group involving key players in the department started development work. It was a Lotus Notes database that described the SHE methods and work processes in offshore projects. The database provided a structure for storing information. This was basically two-dimensional; where the first dimension included the different phases of a project and the second the different areas of SHE in project work (SHE management, technical safety, working environment and environmental care). Later, HVAC was also added to the database. The database included links to steering documents and regulations and to project documents that represented good examples.

Although the needs of a Best practice database were generally accepted, experience showed that it was difficult to motivate senior employees to prioritise the work to develop it. There was, on the other hand, obvious demand for this experience database especially among new employees and external consultants.

In 1998, each discipline responsible was given the responsibility to lead the development of his part of the Best Practice database. Money was allocated to seminars outside the office, where department personnel discussed work practices in order to agree on a Best Practice to be documented in the database. This work was partially successful and in 1999 the first version of the database was about 40% complete. Later, the SHE management, technical safety and working environment parts of the Best Practice database had reached a reasonable level of completion (> 75%).

One year later, the database was expanded to include land-based projects. At the end of 2000, this part of the Best Practice database was only rudimentary developed.

In 1998, all employees were given both read and write access to the database. The idea was that everybody should be allowed to contribute to the Best Practice database, while the discipline responsible should review this input regularly. Experiences showed that this change did not result in adequate improvements in the employees' efforts to update and renew the database. In 2000, the networks were instructed to allocate time for the development and maintenance of the Best Practice database.

There is no detailed account of the total costs between 1998 and 2000 in the development of the Best Practice database. A rough estimate indicates that the total number of man-hours used equals about one man-year (2000 hours).

Figure 6.2 shows the general outline of the database. At the end of 2000, the offshore part consisted of about 200 entries, each of between a few lines to a bit more than an A4 page. Each entry followed a general outline that included purpose, description of method and checklists (if any), advice from those who had performed the activity before and links to reference documents. A total of about 130 study reports and other project documents had been selected to represent Best Practice and were shown as icons in the entries. Further, there were links to other internal databases on steering documents and to the Internet, mainly concerning the legislation in Norway.

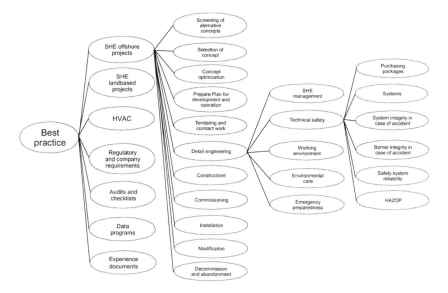

Figure 6.2:   A sample of the tree structure of the Best Practice database.

Development of Technical Specifications and Other Steering Documents

The SHE department was responsible for the development of technical specifications and other steering documents. In the 1980s, technical specifications had been developed for the offshore projects in the areas of technical safety, working environment and environmental care. They represented the company's requirements to the SHE standard in design and were based on the legislation and on own experiences. Department personnel and SHE personnel from operations had participated in the development work. Early in the 1990s, the so-called NORSOK initiative was launched. Its objective was to standardize the requirements for the design of new offshore installations (Kjellén, 2002). Consequently, the company specifications were replaced by common standards for the Norwegian oil industry (NTS, 1996/97). One of the SHE disciplines, technical safety, continued to issue a company specific technical specification as a supplement to the NORSOK standard. It was argued that the NORSOK standard was not specific enough to ensure that contractors actually implemented the necessary design measures. It was also argued that the NORSOK standard could not serve as an adequate experience carrier due to a too long updating cycle. A small group of technical safety specialists from the department and from operations updated the specification regularly.

In 2000, the working environment and environmental care disciplines followed suite. Experiences from recent projects had confirmed the objections of the technical safety people (Kjellén, 2002). The need of a living carrier of experiences from platform operation had become obvious at the same time as the 'political climate' that required compliance with the NORSOK philosophy had changed.

Starting in 1998, new 'generic' technical specifications on SHE for different land-based plants were developed when needed in new projects. They represented the company's core business areas (primary aluminium, fertilizer and petrochemical production) and were all built around the same framework as that of the technical specifications for the offshore industry. Department personnel with relevant experiences and operations personnel were brought in to develop the specifications, and they also went through an extensive hearing before approval. Earlier, each project had developed such specifications without too much interference from others.

One of the recommendations of the quality improvement project was to standardize the design of the projects' SHE programs. In 1999, a new work instruction that defined a 'generic' SHE program for land-based projects was developed. Again, this work instruction broke off a tradition that the development of steering documents on SHE mainly was a concern for the individual project and its SHE manager. A somewhat different approach was selected for the offshore projects. Here, the latest SHE program that had been adequately verified by the basis organisation should represent Best Practice.

Peer Auditing

The quality improvement project claimed that there was an insufficient follow-up of the projects' work in the area of SHE from the basis organisation. The lack of resources at the home office for such follow-up was a concern. It was decided to strengthen the competence of all department personnel in the area of quality auditing in order to make department members in the projects qualified to audit each other. The aims were to improve quality as well as experience transfer and learning among department members. In September 1999, all department members were offered an opportunity to participate in a three days' training on

SHE audits. Participation in an audit was also part of the program for those employees who were inexperienced with audits from before.

Starting in 1999, the department provided input to both the project and the operation divisions' audit plans. The aim was provide for a systematic auditing of project SHE activities, where department personnel from other projects participated as auditors with SHE expertise.

Points of Intervention Seen from the Perspective of Organisational Learning

We will here apply Nonaka's theory of organisational knowledge creation in an analysis of the significance of the different points for intervention (Nonaka, 1994; Hansen *et al.,* 1999; von Krogh *et al.,* 2000). Figure 6.3 models the organisational knowledge creation as a stepwise process, where the steps take place in different 'layers':

1.   Individual knowledge is created through participation in project work. This knowledge is personal and is rooted in actions. It consists of different elements such as cognitive, technical and emotional.
2.   The individual and mainly tacit knowledge is shared within a community of practice. We will here focus on the context for such knowledge exchange that is represented by the formal networks, although other contexts may also be of great significance. Knowledge sharing takes place through rounds of meaningful and time-consuming dialogue.
3.   This sharing of knowledge results in the development of concepts and in the creation of common perspectives. A part of the insight created through these processes is made concrete and explicit and documented in the Best Practice database and other steering documents (specifications, procedures, etc.).
4.   This new knowledge has to be justified as truthful. This is a management task and takes place in the basis organisation. Here standards are used to judge the truthfulness of explicit knowledge, in terms of efficiency and effectiveness and compliance with the legal environment and industry standard.
5.   The new knowledge is then implemented in the basis organisation's knowledge base for use in the projects.

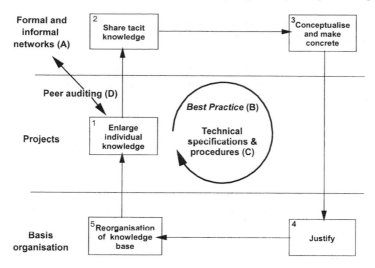

Figure 6.3:   Model of the organisational knowledge creation process. A – D denote the four different points for intervention.

A vital point is to what extent the organisational context enables this knowledge creation process. Nonaka points at several factors that need to be present in an organisation in order for knowledge sharing and the development of common concepts and perspectives to take place:

- The organisational members should be committed to the process by seeing benefits to themselves such as joy and future rewards (better adaptation in the organisation).
- The organisational environment should be trustful and caring and inviting the sharing of experience.
- The personnel that participate in experience sharing should represent an adequate variety of individual experiences (not too little and not too much).
- The organisational environment should be characterized by a moderate degree of fluctuations and chaos to provide for new input and unexpected opportunities.
- There should be overlapping experiences and responsibilities between the organisational members to make it possible to amplify and crystallize knowledge.
- There should be continuity in the relations between organisation members and adequate time for dialogue.

A vital question is whether these enabling factors have been present in the professional networks.

RESULTS

Climate for knowledge sharing and learning

The general co-operative climate and trust within an organisation has bearings on the members' willingness to share experience. The working environment surveys carried out in 1998 and 2000 by the division's Human resources department give an indication of how this co-operative climate has developed in the course of the study period.

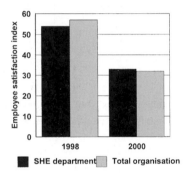

Figure 6.4:   Results of working environment surveys in 1998 and 2000. The employee satisfaction index measures a combination of attitudes, the most important being related to job satisfaction, competence development, compensation and the personnel manager.

Of the different indices of the surveys, the employee satisfaction index is presented here because it represents a summary index of a number of relevant factors. The definition of the index remained unchanged in the course of the surveys. There was a considerable decrease in employee satisfaction between 1998 and 2000, both for the SHE department and for the division as a whole. The single most significant explanation to this decrease is the merger and re-organisation carried out about half a year before the last survey.

The interviews of the members of the department carried out at about the same time as the second survey gave some deeper insight into the results. The department scored low on the employees' feeling of security as a team. There was insufficient interaction and mutual contribution between department members working in the different projects and a low belief in the benefits of co-operation. The employees also felt that there was little support for innovation and creativity. A perceived tension between the project and home office was highlighted. Employees working in the projects felt that they were left on their own with little support from the home office and that there was little concern for their successes or failures. Employees working at the home office, on the other hand, felt that project work had a higher status and gave career advantages and that there was little understanding for their workload and for the significance of their work. Junior members of the department expressed a feeling of insecurity. They felt that the department culture favoured the senior, high-level specialists at the cost of the junior department members and that their knowledge and skills were of limited demand.

The Employees' Evaluations of the Different Measures

Usefulness of different measures: Figure 6.5 shows the respondents' evaluation in February 2001 of the usefulness of different measures. Only the high and very high ratings are shown. The respondents' ratings of the department seminars and the informal contacts within the department and informal and formal eternal contacts are shown as references.

As to the *quality of own work*, department members demonstrated a solid belief in the significance of technical specifications. This answer seems logical, considering the focus in the division on defining contractual binding requirements to SHE and follow-up of the contract. Technical specifications are here important tools (Kjellén, 2002). It is interesting to note that informal contacts also scored high in this respect, and that peer auditing scored relatively low, although the aim was specifically to ensure an adequate quality. The explanation may be that they were relatively infrequent.

The responses showed a similar pattern as to the significance of the different measures for *own competence development*. Here informal contacts scored highest followed by expert networks, external contacts and technical specifications. Junior department members (< 4 years of employment) gave especially the expert networks and the informal contacts high scores.

As to the *feeling of solidarity within the department*, the different scenes for intermingling were given the highest scores. The Best Practice database came out with the lowest score, indicating an insufficient ownership to this measure.

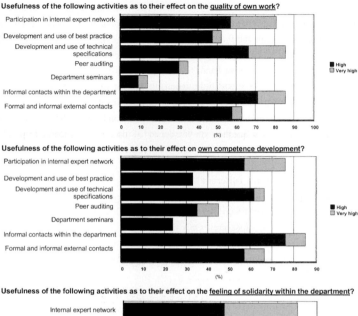

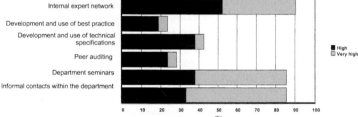

Figure 6.5:   The usefulness of different measures regarding their impact on quality of work, competence development and team spirit (N=21).

Internal expert networks: Internal department networks scored high on all aspects of usefulness. Department members answered that they participated frequent in network meetings (84% participated in more than three of four of the meetings) and that they were active (90%

regarded themselves as active or very active). Network meetings gave adequate possibilities for the members to bring up own areas of concern, whereas more than 50% of the respondents considered that inadequate attention was given to the development of the Best Practice database and new ideas.

Narrative comments in the questionnaires qualified these results further. It was a concern that the network meetings did not allow time for in-depth experience exchange on critical topics and the development of the Best Practice database.

Development and use of the Best Practice database: In general, the Best Practice database development and use was given a low rating on all three aspects of usefulness. Especially the technical safety engineers were critical towards the Best Practice database. These results were confirmed by the responses to a question on frequency of use. Every second respondent indicated that they never or only seldom (about monthly) accessed the database.

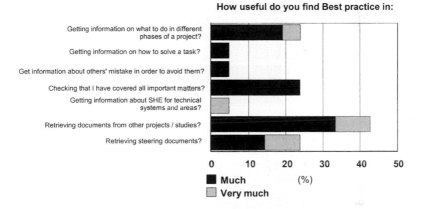

Figure 6.6: Distribution of responses to different alternative statements concerning the usefulness of the Best Practice database (N=21).

Respondents found the mechanical aspects of the Best Practice database (i.e. retrieving of project documents) to be the most useful. Such intended benefits as access to descriptions of appropriate work processes were of little use. These results are underscored by the responses to the question about obstacles to the use of the Best Practice database.

Again, the Best Practice database could not compete with the informal means of accessing information through personal contacts. The quality and completeness of the information in the Best Practice database were also concerns. The fact that respondents did not find what they were looking for was underscored by the narrative comments. The database was considered to be incomplete especially for early project phases and some specialist areas (technical safety, environmental care). Respondents also expressed needs of improving the quality of the database by use of network meetings. The renewal of the database was also a concern. This is underscored by the answers to the question as to how frequently the respondents made input into the database. Nobody made such input more than about once a month. Again, the mechanical aspect of storing project documents dominated the types of inputs.

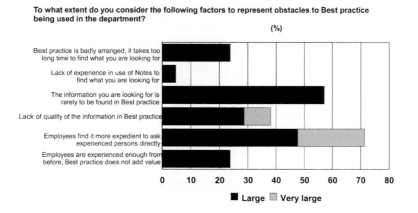

Figure 6.7:   Distributions of responses to questions about the significance of different obstacles to the Best Practice database being used in the department (N=21).

To what extent do you consider the following factors to represent obstacles to Best practice being supplemented with new information?

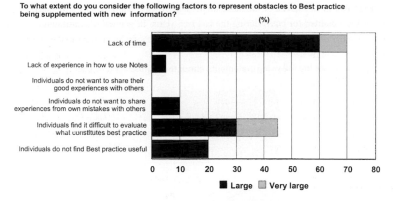

Figure 6.8: Distributions of responses to questions about the significance of different obstacles to the Best Practice database being supplemented with new information by department members (N=21).

Lack of time was stated as the most important reason for not updating the Best Practice database. This was an important reason especially among the senior personnel. (One respondent qualified the evaluation by a narrative characterization 'laziness and individualism'.) Insecurity as to whether experience made through own work really represented best practices was another important obstacle. Respondents expressed needs of a person with the responsibility of coordinating and coaching the other department members in their work to update the Best Practice database.

At a department seminar in March 2001, three different groups addressed questions on the motivation to maintain the Best Practice database, on the incentives to develop and use the Best Practice database and on needs of improvement measures. All three groups concluded that The Best Practice database had a future. They pointed at the deficiencies in the existing version concerning its structure (it was too complicated) and concerning missing information and the variations in quality. They also pointed at the lack of aims and directives for the development of the Best Practice database and at an inadequate incentive structure. All three groups underscored the important role of the networks in the development and quality assurance of the Best Practice database.

Peer auditing: About two thirds of the respondents had at least once either participated in an audit team or been audited (or both) during the last two years. In general, the attitudes towards the usefulness of the audits were rather positive among these respondents, Figure 6.9.

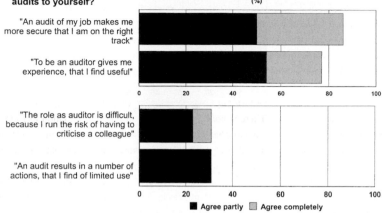

Figure 6.9:   Distributions of the respondents' evaluations of statements concerning the significance of peer audits to themselves (N=14, i.e. all respondents who either had been audited or participated in an audit team during the last two years).

A large majority of the respondents considered the audits to be useful, both as to quality assurance of own work and as to experience transfer. The respondents were somewhat more hesitant to the conflicts embedded in the role as auditor of peers and to the usefulness of the results. The first aspect was highlighted in the narrative comments. Respondents found it difficult to audit peers, when there were close relations. It was stated that audits did not contribute to an open dialogue and to creativity in finding solutions.

DISCUSSION AND CONCLUSIONS

The results of the evaluation paint a rather discouraging picture of the possibilities of capturing individual department members' knowledge and to make it explicitly available as a database of Best Practice to the department as a whole. Department members had not made it into a habit, neither to use the Best Practice database regularly in their work, nor to contribute to it. Of the different uses of the Best Practice database, the 'mechanical' aspect of document storage and retrieval was found most useful.

These findings contrast with the perceived usefulness of works on formal steering documents (especially technical specifications) as well as of less demanding commitments of participation in networks and informal contacts. The significance attributed to technical specifications may be explained by the dominant role that contractor management plays in the project division. Development of technical specifications is instrumental to the possibilities of ensuring that contractors perform an adequate design from a SHE point of view.

The question emerges whether the idea of establishment of an explicit knowledge database of Best Practice is feasible or not. There is not enough evidence from this evaluation to support any general conclusion on this subject. The evaluation gives, however, some insight into the presence (or absence) of what Nonaka calls enabling factors in organisational knowledge creation that may explain the results. Unambiguous results of the survey studies and interviews pointed at a deficiency in employee satisfaction and trust, conditions that according to Nonaka have decisive influence on the provisions for knowledge sharing and learning. The merging and downsizing in 1999 is an obvious explanation to these results. The question also rises whether team spirits and trust in relation to the home office was part of the organisational culture or not. Indications are that the answer is no. The incentive structure, for example, promoted accomplishments, both individual and team-related in the projects at the expense of accomplishments in the long-term development activities of the basis organisation. This is illustrated by the 'project-basis relationship' discussed above.

There were no obvious incentives in contributing to the Best Practice database, especially among senior department members, rather the opposite. Use of the Best Practice database might limit the relatively large amount of freedom that the employees experienced in different project positions in making prioritise concerning own work tasks and in selecting methods of work. Employees saw, on the other hand, obvious advantages in the giving-and-taking relationship of sharing advice in face-to-face contacts between individual department members. Here, the individual decided whom to approach and whether to follow the advice or not.

There were also other constrains in the development of new, explicit knowledge, where the limits in available time plays an important role. Due to downsizing, the span of responsibility and workload of the individual employee had increased. Employees had clear limitations in the time they were able to spend in the work groups that developed the Best Practice database. It followed that there was a lack of continuity and a constant pressure to arrive at a conclusion at the expense of consensus building.

Already from the start in 1996, there had been an emphasis in ensuring ownership among the department members to the database through active participation in its development. It is interesting to note that the rather complicated structure of the database that later has been criticized was the result of a bottom-up approach, i.e. group work with participation of senior and junior department members. The ambition to ensure ownership was for various reasons discussed above such as lack of time and incentives not fully materialized. This may be one explanation for why the database was not used on a routine basis.

Other conditions in the department context may also have had unfavourable effects on the possibilities of knowledge creation. We may, for example, question whether there was sufficient variety represented in the networks to provide for rethinking and innovation. The networks were rather homogenous and stable and there was little emphasize on bringing in personnel representing other areas of experience and ideas. Again, this has to do with priorities due to constrained resources. The organisational culture of emphasizing in-depth expertise and seniority also limited the provisions for bringing in new and unexpected ideas. In general, the organisational culture of the department was conservative and shaped by the responsibilities of interpreting and implementing authority requirements.

Redundancy was also a concern in two of the networks, those of working environment and environmental care. These networks were small and there was limited scope for overlapping of experiences and responsibilities.

It is important, however, not to downplay the significance of the Best Practice database too much. Through the work to develop the database, individual department members have been forced to externalise and benchmark their own work practices. They report improvements in own competence and in the quality of their own work as a result of these processes. There are also reasons to assume that it has strengthened the community-of-practice of SHE engineers by being a concrete output from the formal networks.

To sum up, the results of the evaluation show that several of the measures to improve knowledge sharing and learning, including expertise networks, peer auditing and technical specifications had been relatively successful. The lack of significant progress in the development of a database on Best practice was a concern. The evaluation points at different measures that may increase the propensities for progress. They include:

- Establishment of a department vision that encourages a feeling of joint responsibility for individual members' work in the projects as well as for the department's Best Practice database. This should include an evaluation of the incentive structure for knowledge sharing within the department.
- A clarification and focusing of the scope and aims of the Best Practice database and establishment of criteria for quality assurance (justification) of its contents.
- Further development of the work processes within the networks to develop and justify the Best Practice database.
- Allocation of sufficient time for consensus building in the networks concerning the Best Practice database.
- Review of the size and membership of the networks to provide for adequate variety and redundancy.

REFERENCES

Hansen, M.T., N. Nohria and T. Tierney (1999). What's your strategy for managing knowledge. *Harvard Business Review,* **77**, 106-116.

Kjellén, U. (1998). Adapting the application of risk analysis in offshore platform design to new framework conditions. *Reliability Engineering and System Safety*, **60**, 143-151.

Kjellén, U. (2000). *Prevention of accidents through experience feedback.* Taylor & Francis, London.

Kjellén, U. (2002). Transfer of experience from the users to design to improve safety in offshore oil and gas production. In: *System safety - challenges and pitfalls of intervention* (B. Wilpert and B. Fahlbruch, eds.), pp. 207-224. Elsevier, Oxford.

Nonaka, I. (1994). A dynamic theory of organizational knowledge creation. *Organization Science*, **5**, 14-37.

NTS (1996/97). *NORSOK standards on technical safety, working environment and environmental care.* NORSOK standards No. S-001, S-002 and S-003. Norsk Teknisk Standardisering, Oslo. (http://www.nts.no/norsok/s/)

Von Krogh, G., K. Ichijo and I. Nonaka (2000). *Enabling Knowledge Creation.* Oxford University Press, Oxford.

7

ORGANISATIONAL MEMORY FOR LEARNING FROM OPERATIONAL SURPRISES: REQUIREMENTS AND PITFALLS

Floor Koornneef and Andrew Hale[1]

INTRODUCTION

Learning from operational surprises is about finding solutions to problems that emerge in intended activities. Implementation of these solutions is essential for lessons to be learned (see chapter 2). In order to avoid learning the same lesson twice, if not more times, some form of accessible memory is needed. A person uses her or his brains supported by memory, but for an organisation this is less obvious, because its members sooner or later will leave the organisation, taking their brains with them. Organisations demonstrate to have organisational memory when characteristic features of the organisation remain over time, e.g. as in the case of family-owned corporations. Stein (1995) puts it this way:

> *"The persistence of organisational features suggests that organisations have the means to retain and transmit information from the past to future members of the social system. This capability we might call the organisation's memory."*

He provides in his review of conceptual foundations of organisational memory a working definition:

> *"Organisational Memory is the means by which knowledge from the past is brought to bear on present activities, thus resulting in higher or lower levels of organisational effectiveness". (Stein, 1995)*

---

[1] Delft University of Technology, The Netherlands

In the perspective of learning from operational surprises or incidents, the knowledge that we are interested in, which has to be memorised and reused is about managing operational risks, about unwanted events and situations.

In this chapter, the organisational learning concept described in Chapter 2 is transformed into the SINS-concept (Systematic Incident Notification System) in order to facilitate the discussion of organisational memory for learning from incidents. Initial emphasis is on learning for improving control of operational risks in intended shop floor processes, because it is there where incidents actually occur. However, one single incident occurrence usually relates to many causal factors, among which are a number of factors that cannot directly be changed by the incident process-owning organisation. Characteristically, this applies to incidents related to equipment or other artefacts, e.g. a device that fails to perform as intended. This will usually have been purchased by the user organisation, which has only limited influence on its design or construction. The learning from device failures thus has to take place by the designer, manufacturer or regulator. Thus, organisational memory outside the incident-process owning organisations is discussed also in this chapter.

ORGANISATIONAL MEMORY IN ORGANISATIONAL LEARNING

In this section, we discuss how organisational memory is a pivot in processes of organisational learning.

The Systematic Incident Notification System, see figure 7.2 (Koornneef, 2000), provides an explicit translation of the principles of Organisational Learning discussed in Chapter 2 (see figure 2.2, repeated here as figure 7.1) that can be implemented in organisational systems. The actual notification ('notify' in figure 7.1) goes via a computer interface (SINS-1) that first looks into the organisational memory unit (SINS-2) to see whether the new notification is of an operational surprise that has occurred and been looked into before, and whether there are lessons which have been learned already. If this is the case, then the lesson(s) learned earlier are fed back to the notifying organisational unit for reuse. If not, the surprise represents a new (type of) incident that must be assessed by the learning agency ('inquiry' in figure 7.1).

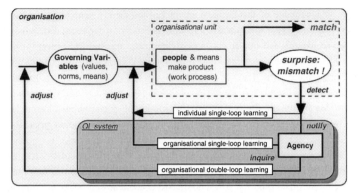

Figure 7.1: Organisational single- and double loop learning (modified: Koornneef, 2000)

From the new surprise, the learning agency (review team) generates lessons to learn, sends these back to the surprise-reporting operational unit ('adjust' in Figure 7.1) and adds the new case to the organisational memory (SINS-2). Initially, the learning agency only acts in the case of incidents that have not been reported before or have occurred under different circumstances and hence contain new opportunities for (modifying) learning. The occurrence of one incident may not be sufficient to make it worthwhile to look for a structural corrective action to prevent recurrence, unless it is readily available. For instance, if the risk associated with the incident is assessed as negligible, than it does not matter whether or not measures are implemented that prevent repetitive occurrence.

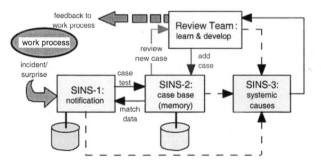

Figure 7.2: The internal configuration of a SINS (Koornneef, 2000)

Yet, frequent recurrence of that incident might be sufficiently annoying to make it worth setting an incidence trigger so that above a specific number of recurrences double loop learning is activated to identify and implement more fundamental measures to prevent recurrence in future. In identifying these systemic causes, the learning agency is supported by the SINS-3 module.

Thus, SINS as a concept fits within Argyris' model of organisational learning with key positions for the learning agency (Review Team) and the organisational memory (SINS-2).

SINS in this form has been implemented in a cardiac surgery unit of a teaching hospital (Koornneef, 2000). The members of the Review Team were also members of the organisational units where the surprises occurred. Thus, notifications could be kept relatively simple, because the Review Team was able to interpret a message in its operational context.

Systemic solutions to the surprise may be more feasible when looked for and implemented in a preceding process phase. Prevention is better than cure, as the doctors say. The cardiac surgery project demonstrated this in several instances (Koornneef, 2000). A nice example is the skin needle, used to keep two edges of the wound skin together in order to close the wound in a neat manner. This needle is not used inside the chest cavity and has therefore not been included on the list of objects used inside the body, which has to be double-checked before closure of the chest cavity. During a quality assurance check of randomly selected patient records, the department manager noticed the skin needle on X-ray pictures made after the operation shortly before the patient would be discharged: although clearly visible, several physicians had not noticed the needle while inspecting the picture for surgical complications. The review team identified two risk control options that could have prevented this incident. A measure to ensure that a physician notices a skin needle on an X-ray picture in the post-surgery phase will require intense supervision of these physicians and will be fallible because of human behaviour features. On the other hand, the alternative measure to include the skin needles in the double-check counting procedure of the nursing staff during surgery before closure of chest cavity, is implemented and maintained easily. This small change in the counting protocol is in itself a minimal modification of relevant organisational memory. Incidents that occur in one step in a process must therefore be available to (and preferably referred actively to) the review team which considers earlier steps in the process or steps and processes which deliver the essential requirements to that primary process.

Resources for the learning agency were limited in the hospital case as its members already had a high workload. This is often likely to be the case. For this reason, the organisational memory is of particular importance. Lessons, once learned, need to be stored and become available when appropriate. The more lessons learned, the simpler the notification messages (using SINS-1) can be if appropriate cases in memory are retrieved effectively and efficiently.

Handling repetitive surprise notifications

SINS makes it possible to handle repetitive, no- and low-risk notifications efficiently. It does this by ensuring - through the learning agency - that the first notification of a particular occurrence is stored in organisational memory (SINS-2) together with any lesson(s) learned. Subsequent notifications of similar occurrences are recognised by SINS-2, e.g. by means of Case-Based Reasoning technology (Aamodt and Plaza, 1994; Weber *et al.*, 2000), and their frequency is updated in memory. In this way, the incidence of a certain type of surprise is determined without further involvement of the learning agency.

Case-Based Reasoning technology recognises keywords and aspects that are coded early in the notification process and matches these to clusters in its memory. As soon as the match is good enough, it can stop the notification by putting on the screen its own generic description and asking the notifier if this is what he/she is trying to describe. The notifier can then click on 'agree' and end the notification, or modify some or all of the details of the generic description to indicate that what has happened is a variant on the generic incident, or is quite radically different.

The learning agency classifies each new surprise (and countermeasures) according to scenario type and risk acceptability before adding it to memory. For each type of incident, the agency also determines an incidence value, above which underlying causal factors need to be resolved (via double loop learning) in order to prevent recurrence: the incidence trigger level *IncT*. The use of the incidence triggers allows investigative resources to be prioritised. When the event frequency reaches the *IncT* value, the learning agency is warned and can invoke a double-loop learning process. To support this, the learning agency may use an expert system on organisational causal factors (SINS-3). An example described by Koornneef (2000) is an incident review system that uses the Management Oversight and Risk Tree (MORT; (Frei *et al.*, 2002; Johnson, 1973) as a knowledge model in an expert system where experts have added diagnostic rules in order to generate a diagnostic report indicating management system problems underlying the incident. Operational problems relate to inadequate functioning of hardware, procedures or people, which in turn link to systemic management system factors such as policy and its implementation, available resources, process design including requirement specification and operating procedures, assurance of operational readiness, inspection, maintenance, supervision, selection and training.

The accumulation of data in SINS includes the incidence of a particular type of incident (linked to *IncT*), as well as of the incidence of the causal codes assigned by the Review Team to systemic conditions in validated cases. Causal codes indicate categories of systemic causal factors or 'root causes', e.g. common mode organisational factors, or links to organisational systems delivering resources or procedures to manage risks (Bellamy *et al.*, 1999). The cumulated causal factor data can be analysed to highlight weaknesses in the risk management system. They might be used to generate root cause profiles similar to those in PRISMA (Vuuren, 1998) and TRIPOD (Groeneweg, 1998). Root cause profiles help to prioritise focal areas for improvement of the risk management system of an organisation, e.g. the maintenance system or on-the-job training programs. Those causal factors that fall within the span of control of management of the next higher organisational system level, form work process data that can be turned into lessons that might be shared between organisational units. For example, a design flaw in software for blood bank computer systems so that requested blood type is not verified against the patient's blood type, can result in fatal administration of wrong blood to a patient. The supplier can provide a simple software patch to prevent future mishaps. This lesson should be shared without delay with other blood banks that use the same software. Any remaining common-mode causal factors that lie at even higher levels or in other dimensions need to be extracted from this analysis and fed to the appropriate place in the organisation. For instance, a structural shortage of qualified staff noted in several incidents may require that the human resources manager decides to change strategies on recruiting, selection and training of new personnel.

The initial notification can be kept simple if the learning agency consists of members with experience of the work in the units which are generating the surprises. Such agents then only need half a word to understand the surprise message. For instance, an experienced nurse, as a member of the learning agency, knows the operational settings in the Medium Care Unit on a day-to-day basis and does not need these to be reported when an incident occurs here, unless they were extraordinary. In other words, the learning agency applies informed process models to the data model underlying each message that arrives for review. A basic sequential business process model such as that described in chapter 2 provides an effective context-descriptor in a surprise notification, also for situations where different process phases occur simultaneously in one geographical area. The details of what happens in these process phases can be kept in the organisational memory - or in the human memories of the review team members.

EXTERNAL MEMORY IN ORGANISATIONAL LEARNING

Although above the focus was on learning from operational surprises in order to improve management of operational risks, not all organisational learning can be achieved by the incident process-owning organisation. Hence, must be organised elsewhere, which will only have an indirect effect on improving the process in which the incident occurred. An example will help to illustrate this.

Health care example: Learning from product-related incidents

We will use as the example a health care provider who runs into a problem when trying to deploy a medical device. It concerns a hospital ward, where a functional device failure, such as a malfunction of a defibrillator (a device to control heartbeat rhythm), can disrupt the intended care delivery process. Several relevant managers need to be identified here who may learn distinct lessons.

User management

The unit management will try to restore process control in any case, e.g. by using a back-up device to save the patient and end the operational emergency mode. The health care institution in which the health care process is embedded is responsible for this service and, therefore, management is focussed on the functioning of the care delivery units. The hospital department manager who supervises the organisational units within that department, will need to learn lessons about training, inspection and may need to change the purchasing policy if the defibrillator failure is the latest in a series. These lessons aim at improving operational readiness and control by the care delivery unit in the case of some failure of the specific kind of device. However, this operational process manager is not embedded in the manufacturing system and so cannot contribute to learning there. He is only able to recognise a recurring device problem once it has happened several times in his own hospital. It should preferably happen earlier at a global level. Systemic solutions to such operational device problems include the design of the device and market approval regimes. The manufacturing and marketing of medical devices is not an objective of the individual health care institution. Therefore, to secure learning about that aspect, the notification process must reach outside to the organisations responsible for the design, manufacture and selling of medical devices. User managements also need accessible memories in which incidents relevant to them are stored so that lessons can be shared among similar use organisations.

Supplier management

Operational problems with devices may indicate a need to modify the device or to withdraw the product from the market, especially in the case of critical devices (e.g. classes II(ab) and III in European as well as in the Food and Drug Administration's (FDA) classification schemes (EC, 1993, FDA, 1997)[2].

The actual extent and severity of problems with a particular device could be assessed if the incidence rates were known, but for individual institutions these are usually low. Combining incidence data on a much larger scale, e.g. worldwide, amplifies the signals coming from device-related operational surprises so that the signals concerning a specific device can be recognised earlier, leading to faster action. Thus, the question arises where we might find an executive manager who can resolve device-related problems systemically?

An evident answer would be: with the device manufacturer. He can decide to alert the user to take extra care, or avoid certain use conditions. He can also modify, if not withdraw, a device in response to signals coming in directly through user reporting systems or indirectly through other channels. Manufacturers can influence user operations by means of changes to their products or their recommended use. However, if this were done device by device, the users would need to report to and interact with each device manufacturer. A far greater problem is that each relevant manufacturer can impose filters for incoming signals. The BS-cc heart valve case (CEC, 1990) illustrates such manufacturer-designed and owned system with a feature to absorb or filter out surprises rather than recognising them. In that case the reporting system collected numerous and revealing statistics on valve failures, but responded only with "Dear doctor" letters to calm down health care professionals. This gave it a notorious reputation in the BS-cc heart valve case. Choosing the manufacturer as owner of the surprise information system places heavy reliance on the integrity of individual companies, which is not a priori justified, as Pfizer showed in the BS-cc case [see also the proven deception of the public by the tobacco industries (Glantz *et al.*, 1998)].

Manufacturers whilst needing to market satisfactory products (assuming a competitive marketplace) want to minimise the number of false alarms. False alarms create wasteful activity as well as undermining consumer confidence. However, it is generally true that the more one reduces false alarms, the higher becomes the miss-rate (i.e. increased numbers of false rejects).

---

[2] Class II relates to medical systems with medium application risks, e.g. minimally invasive devices, diagnostic devices, implantable devices, and devices with influence on blood or cells. Class III concerns high application risks, e.g. implantable and long-term surgical invasive devices with direct contact to the heart, the central circulatory system or the central nervous system.

Regulator management

If we want to avoid this pitfall and improve the chance of the medical device reporting system working, another option as learning agency would be a governmental regulator, for example the FDA. Such a regulating body imposes (pre-) market approval rules for medical devices. A manufacturer must comply with these rules in order to stay in business. The linkage between this regulatory manager and the users who experience problems with the application of devices could be designed as a surprise notification system that is accessible by anyone who uses a medical device. Although this manager will also design filters for this surprise monitoring system, e.g. to avoid overload or filter out device-unrelated problems, these will be independent from device producing manufacturers and may be questioned elsewhere, e.g. by a parliament. An intervention by this regulating manager can directly affect user processes by a persistent message, e.g. a "STOP using this device" ruling. The US MedWatch program (FDA, 2002) is such a system, which enables any user of a medical device to notify problems with using a medical device directly through Internet.

The memory part of this surprise monitoring system consists of a (raw) database (MAUDE - (CDRH, 2002), accessible in the public domain via Internet. For the following reasons we have to conclude that, at present, it is less than adequate. Submitted data are not verified and it is very difficult for users to retrieve useful data from the database. In order to be useful for learning lessons, such a low-threshold notification database should contain clues about the observed versus expected behaviour of the device, and regarding the conditions of operational use, especially whether or not the device is used under real-time risk control by the health care provider. For instance, one would expect that an implanted cardiac pacemaker is in active mode, thus not in read-out mode, when the bearer visits the health care institution for a periodical check. Furthermore, the normal operational use of such a device by a bearer is outside the controlled environment of a hospital.

A regulatory body as regulatory manager has a great interest in a well-designed sentinel surprise monitoring system that has effective 'early warning' features without creating too many false alarms. Yet, current incident databases are hardly effective as memory in organisational learning due to limitations in retrieval techniques (Johnson, 2000), but moreover because the 'loss-of-context' in the incident data cannot be compensated efficiently in retrospect (see also Chapter 2).

Discussion: multi-dimensional organisational memories

Learning from product-related incidents needs to be organised *beyond* the user and supplier organisations to enable identification of use problems that come with a class of products. Table 7.1 illustrates the complexity of organisational learning processes at different levels of learning from product-related incidents, in which memories are required to retain relevant lessons for reuse.

Table 7.1: Complexity of organisational learning processes at different levels (Koornneef and Voges, 2002)

level	*user - products relationships*	*ratio*	*lessons on*
User	1 use process : many products	**1 : m** ➔	operational risk management options & process control modifications
Supplier	many use processes : 1 product	**n : 1** ➔	device design requirements (improvement of hardware & software, incl. protocols
Regulator	many use processes : many products	**n : m** ➔	market approval (withdrawal, adjustment of regulation)

At user level, the health care provider utilises many different devices for distinct tasks and applications in the care delivery process. A pacemaker that is found to be in the wrong mode of functioning would be just one of these many devices, although a critical one. At supplier level, a manufacturer has a limited range of products of which units are sold to users in many distinct organisations running a wide variety of use processes. However, at product level, a manufacturer usually has competitors. Thus, not only needs a manufacturer to learn from incidents in use processes that relate to his product, but the lesson may also need to be learned by his competitors in order to prevent specific types of product-related incidents. The regulator is facing many products from many manufacturers that are applied in a wide variety of use processes. To the regulator, the pacemaker represents just one class of safety-critical products. A problem identified in relation to a device of a specific brand may be common for all makes. The regulator has interest in knowing this timely.

Requirements for a data model for (initial) incident notification are not the same for these distinct levels. Yet, all are dependent on the user for the initial notification because problems occur in that user process. The content requirements for the organisational memory depend on the type of lessons that can be learned by a learning agency for a specific set of shop floor operations.

The sharing of cases stored in organisational memory of the user organisation thus might be limited to issues of process control in organisational units or other organisations that run similar operations. User organisation are also interested in sharing learning about shared management system factors within an organisation or problems related to the use of particular technological systems. The learning of these sorts of lessons is dependent on being able to check or verify whether the conditions of use in each unit or organisation are sufficiently similar. When we move to the learning which the manufacturer or supplier needs to realise, the relevance of the conditions of use or operational context of the incident is reduced, but not entirely lacking. For instance, application problems may be due to inadequate design assumptions about user behaviour or operational settings, even when the product functions to specification. Therefore, data that is relevant to the supplier includes application and use contexts at a more generic level, relevant to identify DOs and DON'Ts requirements for the product. For example, it does not matter whether a cardiac pacemaker is switching modes (active to standby) unintentionally due to electromagnetic impulses in a anti-shoplifting gate or while searching for treasures with a metal detector: electromagnetic fields can be anywhere without warning! The point is that - by design - such mode switching must be detected and recovered from. Suppliers of one type of apparatus need to be alerted to think of all use scenarios where such switching can take place. Finally, such lessons may well deserve to be shared even more widely, e.g. manufacturers of all classes of products, which are sensitive to mode, switches.

MEMORY FOR ORGANISATIONS: A CLOSER VIEW

Organisational memory is manifest in many different forms. For instance, experience built-up about how to fine-tune specific processes might be captured in updated operating procedures and training programs or in models governing process control systems, each form being part of the organisation's memory.

Organisational memory needs to be detached from individuals, who sooner or later leave, in order to retain the experience and built-up knowledge in the longer term. The tacit memory in the head of individual members of an organisation may be sufficient in the short term for learning from incidents. But in the longer term, more tangible forms are needed in order to share and disseminate lessons learned, e.g. by means of a 'lessons learned' journal such as that in the CIRAS system (CIRAS, 2002, see also chapter 10). Retrieval of lessons from such a memory must be made easy. This means at least that memory must be easily accessible and must exist where it makes sense for the user to look.

Numerous concepts exist about what Organisational Memory is, what it contains and how it might function, see Stein (1995) for an overview, and also Chapter 5 in this volume. Heijst *et al.* (1996) point out that the most suitable form and configuration of corporate memory depends on the purpose and how it will be used.

Bannon and Kuutti (1996) argue that organisational memory, when conceived as a passive depository of data, knowledge and information is doomed to remain of little use. This is because users still have to interpret the data retrieved from the organisational memory, but often will have no cues about time, originator, context, etc., aspects which are essential to understand and use the information. They stress the active and constructive aspect of remembering in human activity at both the personal and collective levels. This has implications for organisational memory, triggering us to focus on the way information is generated and stored, and subsequently is interpreted and understood by other people in other settings at other times.

Konda *et al.* (1992) argue for a 'shared memory' that has two forms: vertical and horizontal. The 'vertical' shared memory contains the body of knowledge specific for one professional group or discipline. The 'horizontal' shared memory contains knowledge with a consensus and meaning shared by different professional groups participating in a particular design project, use process or operation (Bannon and Kuutti, 1996). An example of what can be stored in this horizontal memory is the aims of and approach lying behind the process. The point here is that contextuality of information in memory is a necessity.

Bannon and Kuutti refer to (Bartlett, 1932) to emphasise that it is essential, when looking into the utilisation of organisational memory, to consider processes of memorising and remembering:

> *"But if remembering takes place in a different activity than the one where material has been stored, the material will be reinterpreted with respect to the new object of activity, and there is no automatic guarantee that the material is relevant anymore in the same way than it was in the context of storing it." (Bannon and Kuutti, 1996)*

They observe that the attention for the contextuality of contents in organisational memory has been neglected in many studies.

Location of organisational memories

As described above (Table 7.1), the lessons that can be learned from incidents differ for different stakeholders. What does this say about requirements for a relevant organisational memory? First, that these requirements vary for different groups of users. The user experiences the operational surprise, and will not notify it unless there is an incentive, such as seeing that the organisation is indeed learning through a visible learning agency. A device-related problem will only be notified by the user to a system of which its organisational memory is accessible and independent from device manufacturers. This memory must link to all relevant surprise generating work processes, otherwise, each of the user organisations would need to maintain its own device-related memory and would not see the value of investing such resources.

Second, the user memory must be designed and embedded in the organisational learning processes, so that lessons learned come back (eventually) to the organisational unit in order to implement the learning and to maintain the incentive to users to keep notifying operational surprises in future. In the example above, the original notifying organisational unit may be involved in three distinct processes of organisational learning, two of which having organisational memory outside the user's organisational system (i.e. within the hospital, and those linked with manufacturers respectively with a regulatory body). The notifying user does not need to be burdened by these different learning processes, as long as the user can notify directly via one of these organisational learning systems, as is the case in the MedWatch program (FDA, 2002). The learning agency and the design of the organisational memory must enable the system to route bits of a notification message towards organisational learning systems that reside outside the user's organisation.

Memory contents

Lessons learned in one organisational unit may be valuable for other units, or comparable entities in other organisations (e.g. between all cardiac surgery units in The Netherlands), but requirements for message contents are far from obvious. The coding of the relevant situational context of the surprise(s) that resulted in the learning of a particular lesson needs closer attention in order to avoid trying to reuse a lesson in an unsuitable situation. This issue is crucial in the design of the organisational memory as discussed above, and the projects up to now only give initial indicators of how this must be done (Koornneef, 2000). Weber *et al.* (2000) distinguish eight ways for categorising lessons learned systems to help identify an adequate design methodology for intelligent lessons learned systems, but also recognise many open issues regarding contents, which still have to be sorted out.

Besides the reuse and sharing of lessons between comparable organisational units, there is also the transmission of surprise data from one organisational level upwards within and beyond the organisation, e.g. to corporate, national or international level. The goals of organisational learning from such data can be twofold. Firstly, to achieve high-level control or adjustment of lower-level operations or achievement of learning. In this case the purpose of the upstream flows of surprise data is just to generate performance indicators for use in setting priorities in policy and resource allocation. Secondly, to disseminate lessons faster than is possible horizontally (e.g. about product-related hazards where manufacturers and regulators need to be reached). For each of these purposes, the design of data filters in order to put messages across rather, than just data is far from a trivial problem, and is one that deserves further exploration.

CONCLUSION

In this chapter, we have seen that distinct feedback loops exist through different processes of organisational learning, each organisational learning process involving at least one learning agency. Several issues are worthy of attention. Firstly, a single incident may contain lessons for different relevant managements in different organisational systems. Secondly, device-related incidents may occur in different user systems in parallel: one single (related) user organisation might not recognise rapidly enough a problem that is related to the use of a specific device. Thirdly, distinct relevant managers can only learn lessons that match their specific spans of (risk) control. Fourthly, device-related operational surprises *always* occur in user processes, thus, calling for notifications from this level. However, resulting processes of organisational learning may (need to) go well beyond the operational-surprise-generating user organisation. Fifthly, organisational memory should match the organisation where lessons are (to be) learned. Such memory will differ for distinct relevant managers. A key function of organisational memory is to retain lessons once learned in such a way that they will be reused if an opportunity for learning the same lesson again emerges. Sixthly, the time needed to learn lessons through distinct organisational learning processes will typically be different. Users need to learn lessons in real-time; designers may only be able to learn for the next generation of their product. Lastly, organisational memory must be fed and maintained in order to make it part of active organisational learning processes for improved management of operational risks.

REFERENCES

Aamodt, A. and E. Plaza (1994). Case-based reasoning: Foundational issues, methodological variations, and system approaches. *Artificial Intelligence Communications,* **7**. www.iiia.csic.es/People/enric/AICom.html

Bannon, L. J. and K. Kuutti (1996). Shifting perspectives on organizational memory: From storage to active remembering. In: *Proceedings of the 29th IEEE HICSS Vol. III, Information Systems - Collaboration Systems and Technology,* pp. 156-167. IEEE Computer Society Press, Washington.

Bartlett, F. C. (1932) *Remembering.* Cambridge University Press, Cambridge, UK.

Bellamy, L. J., I. A. Papazoglou, A. R. Hale, O. N. Aneziris, B. J. M. Ale, M. I. Morris and J. I. H. Oh (1999). *I-Risk: Development of an integrated technical and management risk control and monitoring methodology for managing and quantifying on-site and off-site risks.* Ministry of Social Affairs and Employment, The Hague, The Netherlands.

CDRH (2002). *Manufacturer and User Facility Device Experience Database (MAUDE).* Center for Devices and Radiological Health, Rockville MD, USA. www.fda.gov/cdrh/maude.html

CEC (1990). *The Björk-Shiley heart valve: "earn as you learn". Shiley Inc.'s breach of the honor system and FDA's failure in medical devices regulation.* US GPO 26-766. Committee on Energy and Commerce, U.S. House of Representatives, Washington DC, USA.

CIRAS (2002). *Confidential Incident Reporting & Analysis System - How the system works.* CIRAS, UK. www.ciras.org.uk/system_works.html

EC (1993). *MDD - European Directive on Medical Devices.* 93/42/EEC. http://europa.eu.int/comm/enterprise/medical_devices/communitywidelegalframework.htm

FDA (1997). *Federal Food, Drug, and Cosmetic Act as Amended by the FDA Modernization Act of 1997.* www.fda.gov/opacom/laws/fdcact/fdctoc.htm

FDA (2002). *MedWatch: The FDA Safety Information and Adverse Event Reporting Program.* FDA, Washington DC, USA. www.fda.gov/medwatch

Frei, R., J. Kingston, F. Koornneef and P. Schallier (2002). *NRI MORT User's Manual (generic edition).* NRI Foundation, Delft, The Netherland. www.nri.eu.com/NRI1.pdf

Glantz, S. A., J. Slade, L. A. Bero, P. Hanauer and D. E. Barnes (eds.) (1998). *The cigarette papers.* The University of California Press, San Francisco, USA.

Groeneweg, J. (1998). *Controlling the controllable: the management of safety.* DSWO Press, Leiden, The Netherlands.

Heijst, G. v., R. v. d. Spek and E. Kruizinga (1996). Organizing corporate memories. In: Procs. of the *Tenth Banff Workshop on Knowledge Acquisition for Knowledge Based Systems* Eds.). http://ksi.cpsc.ucalgary.ca/KAW/KAW96/vanheijst/HTMLDOC.html

Johnson, C. (2000). Software support for incident reporting systems in safety-critical applications. In: Proceedings of the *Safecomp 2000* (F. Koornneef and M. v. d. Meulen, eds.). Springer, Rotterdam.

Johnson, W. G. (1973). *MORT: The Management Oversight & Risk Tree.* SAN 821-2. US AEC, Washington DC, USA. www.nri.eu.com/san8212.htm

Konda, S., I. Monarch, P. Sargent and E. Subrahmanian (1992). Shared memory in design: a unifying theme for research and practice. *Research in Engineering Design,* **4**, 23-42

Koornneef, F. (2000). *Organised learning from small-scale incidents.* Delft University Press, Delft, The Netherlands.
www.library.tudelft.nl/dissertations/diss_html_2000/tpm_koornneef_20000926.html

Koornneef, F. and U. Voges (2002). Programmable electronic medical systems - Related risks and learning from accidents. *Health Informatics Journal,* **8**, 78-87

Stein, E. (1995). Organizational memory: Review of concepts and recommendations for management. *International Journal of Information Management,* **15**, 17-32

Vuuren, W. v. (1998). *Organisational failure - an exploratory study in the steel industry and the medical domain.* TU Eindhoven, Eindhoven, The Netherlands.

Weber, R., D. W. Aha and I. Becerra-Fernandez (2000). Categorizing intelligent lessons learned systems. In: *Intelligent Lessons Learned Systems: Papers from the AAAI Workshop* (D. W. Aha and R. Weber, eds.), AAAI Press, Menlo Park, CA, USA. www.aic.nrl.navy.mil/AAAI00-ILLS-Workshop/papers/RWeber00.ps

8

COMPUTER-BASED KNOWLEDGE SHARING – SUCCESSFUL IN "OPEN" ORGANISATIONS

Ruth Marzi[1]

INTRODUCTION

The diversity of powerful computer based tools to support – mainly intra-organizational – learning, is large. They range from project management to decision support and are varying in degree and area of promoting personal and organizational learning. The parenthesis embracing all these efforts is subsumed under the term knowledge management.

Information technology (IT) solutions can only work, if the culture in an enterprise is such that it nurtures the willingness to learn – even from mistakes, if openness is lived in the company – throughout all levels. Introducing any of these tools in a company and using them effectively means that the organizational structure changes with regard to the knowledge within this enterprise.

Are expert systems the solution to preventing the loss of individual and organizational knowledge? Yes and no. Depending on how they are built they might even lead to a decrease of competence of individuals. To counter this process and even reverse it, the multi-agent

---

[1] Berlin University of Technology, Germany

system ComPASS[2] was implemented, and an instrument was developed to measure competence enhancement.

KNOWLEDGE MANAGEMENT

Knowledge management (KM) is the systematic acquisition, provision and handling of knowledge within an organizational context in order to ensure increase and continuity of knowledge or conformity with company standards in the conduct of processes. Also, sometimes expertise within an enterprise is not known and therefore is not made use of. As a consequence, work is done inefficiently and ineffectively (inventing the wheel anew, redundancy in work, no documentation of – positive and negative – experiences) (Eppler, 1999). KM-Systems are computer-based systems, which help an organization provide the infrastructure for successful knowledge management. These are usually integral systems, attached to the management level. Even though, also specialized software-systems can make their contribution (stand alone) by systematically supporting learning of individuals and groups.

The introduction of knowledge management is usually economically motivated. But in order to reach the positive long-term effects of an implemented knowledge management, the climate for organizational learning has to be established beforehand. Changes always cause fear and unsettlement in the personnel as to their own potential and flexibility to follow those changes. For the management it means a change in the leadership.

In differentiating between levels of organizational learning, i.e. system-oriented vs. detail solutions, there is a tendency to leave out the theoretical approach in order to establish short term gains by directly implementing detail solutions. System-oriented solutions affect large parts or the entire organization through the introduction of management tools or common access to software for planning and controlling purposes and take some time to be utilized and accepted by all. Detail solutions on the other hand, such as specific computer based support systems, can be installed on a group or individual level. And even though the disciplines concerned with learning of individuals do not always have the optimal solution, unfortunately they are often neglected in implementing solutions to increase the knowledge of an organization. A challenge for KM-systems is definitely its trans-disciplinary impact. In organizational terms disciplines are usually organizational units dealing with a certain aspect of

---

[2] Research on ComPASS was funded by the German Research Association (DFG), funding number Ti 188/6-1 and Ti 188/6-3.

the enterprise, such as controlling, marketing, R&D, a.s.f. All theses "disciplines" with their different terminology, professional background and knowledge have to be able to participate in the advantages of the KM and therefore need an interface tuned to them. This can lead to questioning the boundaries of these disciplines (Döhling-Wölm, 2000).

The issues concerning the establishment of a climate for KM in an organizational unit shall not be addressed here, but such a climate is a necessary prerequisite for its success. In the following, an example of support of KM through IT is presented. Whether, where and when it is useful to employ such an IT-system has to be decided individually on a case-by-case basis by the organization introducing and using it. Such systems are no guarantee for successful KM, but can form a solid base for putting into practice the management commitment to KM. In other words, technology cannot unfold its potential, if it is not culturally embedded in the organization.

Knowledge management also requires the willingness by the personnel to take in information, generate knowledge and accommodate behaviour accordingly. So central for the use of IT in learning organizations are the questions: "Which properties are needed in a KM-Systems to successfully implement KM-Systems?" and "Which processes can be successfully managed by KM-Systems?" What are the advantages of using IT-systems as a means for successful KM? The main advantage of computer based tools such decision support systems is that they allow the gathering of data in a structured way, which helps people organize their knowledge and focus on the relevant issues. They prevent mistakes that might otherwise happen through neglect or misjudgement of their importance. They also allow the evaluation of success and failure in qualitative and quantitative ways.

In addition to learning, a collective organizational memory is needed. A painful experience alone does not ensure a change of behaviour, remembering such an experience can even lead to an avoidance of a situation and thereby a repetition of an erroneous decision. Whereas the individual memory is inside and immanent of the system, the organizational memory has to be (artificially) created and resides outside those contributing their knowledge (Döhling-Wölm, 2000). Such an organizational memory is easily supported by IT, such as data-bases, information systems or work-flow management-systems (WFMS). It is far more difficult to transform this information into knowledge to be used by all. Some of the advantages of decision support are described in more detail in the following.

Decision support systems (DSS) help the user make decisions by providing the underlying information. They may also present the possible decisions, their advantages, disadvantages and consequences. The final decision, which proposed alternative to choose, remains with the user.

The question of what distinguishes data from information from knowledge has been tackled in various disciplines ranging from social sciences to cognitive sciences to computer science. In the following the distinction will be made according to (Wilke, 1998), where the basic operations on data are coded observations, on information systemically relevant data and on knowledge the integration of information into contexts of experience. The restrictions on data are numbers, language, texts and images; those on information its relativity concerning the system under consideration, and on knowledge the "community of practice".

The challenge regarding knowledge is its transfer to other units (persons, organizations). The management of knowledge is therefore the steerage of the communication process to generate knowledge and to ensure safeguard and development of the knowledge base. (In this context a knowledge base is not necessarily implemented in an expert-system. There, the knowledge base contains - permanent - knowledge of an expert and - temporary - knowledge gained in a specific problem solving situation.)

The more a software system requires active involvement by the user, the more probable is successful learning and behavioural change. Furthermore, with increasing degree of interaction (with learning instances) the problem-solving competence and team-oriented learning increase, too (Döhling-Wölm, 2000). Expert systems offer a high potential to support such evolutions, since they are powerful tools, which may even contain the knowledge of more than one expert. They make knowledge available to all, overcoming the bottleneck of experts as a resource. They ensure the continuity of knowledge in the enterprise, acquaint new personnel with procedures, thereby following a strategy of tackling problems adhering to company policies. Furthermore, they present problem solving approaches, which may be new to users. The "dark side" of expert systems is the huge effort to build, maintain and update them in order to ensure consistency and completeness of knowledge. Their limits are every-day knowledge and emotional intelligence. Being complex systems they are also difficult to understand. If users rely on the solutions given by these systems, they might lose competence by forgetting procedures. Expert systems are only suitable for a restricted range of domains: the knowledge has to be complex, but not too complex and it may not change too quickly.

So in accordance with previously cited statements it is important to leave decisions to users. Then competence is not lost, because the user is required to "think along" the problem solving path. If strategies are presented - and explained, which are new to the user, his competence increases over time, instantiating learning by doing, i.e. training on the job.

In the following a solution to the problems described here is presented. The system ComPASS is based on a multi-agent-architecture processing expert knowledge on fault diagnosis in CNC-machine-tools.

THE SYSTEM ComPASS

The Competence Promoting Multi-Agent Support System with Speech Output (ComPASS) was developed within a project in the application area of fault diagnosis of CNC machine-tools (Marzi and Timpe, 2000). A multi-agent structure was chosen, because it offers the possibility to distribute the knowledge about the problem solving process among the agents (Bechtolsheim, 1993). Since complex facts are memorised better, if presented in a multi-perspective view (Paivio, 1971; Bergmann, 1994), each agent of the system supplies the user with a different view on the problem.

ComPASS is a decision support system, which helps CNC-machine-operators diagnose faults and supports them in their repair job. Especially in small and medium sized enterprises (SMEs), operators on the shop floor are not sufficiently qualified for this task, when it comes to difficult problems. So, service personnel – often provided by the manufacturer of the machine – is called in to repair the fault, which results in a prolonged standstill of the machine. In order to qualify the operators, IT-support is needed, not for the routine daily repair jobs, but for serious, rarely occurring, i.e. difficult problems.

"Difficult" is defined in this context as:
1.   the solution of the fault finding process is difficult to obtain
2.   the solution of the fault finding process is not known to the operator.

A support system has to aid both, the fault finding process and the repair. If the operator is merely presented with the solutions and instructions on what to do, the problem is solved, but the operator has not gained any knowledge. So, when the fault reoccurs, the operator has to engage the help of the system again. A better decision support system helps the user increase his competence, while encountering the fault for the first time. This is attempted by intensive interaction during the fault finding process. The decision on which strategy(ies) to employ and which steps to take are left to the user. He is repeatedly asked to input or select items, thereby being forced to "think along".

Goal of ComPASS

The goal of the project described was to develop a system, which supports the user and enhances his competence. Therefore several features had to be defined:

- To obtain competence enhancement, the system should offer solutions from different perspectives for the malfunction encountered and leave the decision, which hypothesis to pursue, to the user.
- To support the user's decisions, the system should offer different opinions about a problem. This requirement is deduced from the fact that different people respond differently to information presented. So, different opinions about the problem may be based on different views on the machine (see below). Also, often there is more than one way to solve a specific problem.
- To overcome problems with unknown malfunctions or machine state changes during work, the user should be able to enlarge or modify the system himself.
- To enhance the operator's skills, he should be motivated to an intense interaction with the machine.
- To support the operator in his personal fault finding procedure, the system should assist these individual diagnostic strategies and, if desired, offer additional diagnostic strategies (John and Timpe, 2000).

Architecture of ComPASS

ComPASS consists of two deliberative, adaptive, loosely coupled agents, which each use their own inference strategies (fig. 8.1). The user determines, which agent to utilize for the diagnosis. Furthermore, two supporting agents were added, one for protocoling and one for tutoring novices (Marzi *et al.*, 2002).

The employed multi-agent structure enables the system to present the solution and the problem solving path from different views. The Vertical Agent (VA) offers an in-depth view on the machine. Machine components are decomposed hierarchically into parts and subparts. The so-called Horizontal Agent (HA) on the other hand sees the machine as joint parts. It has knowledge about the connections between different parts of the machine. So, e.g. the paths of signals (hydraulic, pneumatic, electrical or mechanical) may be followed through the whole machine. This agent searches for the source of the fault along the functional dependencies between component parts and assemblies of the machine (Marzi and John 2001a).

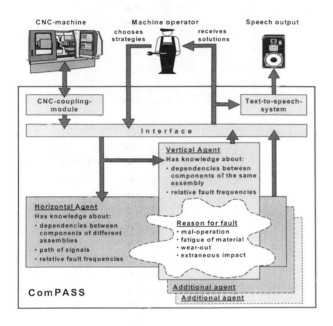

Figure 8.1: ComPASS-architecture (Marzi and John, 2001b)

The agents employ human diagnostic strategies during trouble-shooting, selected from well-known, empirically determined strategies (Konradt, 1995), such as "path of signal" and "relative frequency" for the horizontal agent and "relative frequency" for the vertical agent. These different strategies are supported by the different agents, some by the VA, others by the HA, and some even by both. Depending on which strategy a user chooses, he is supported accordingly. The solution is presented together with advice on how to remedy the problem. Additionally, the system may suggest to the user a different strategy which might lead to a faster solution (as one aspect of learning).

User interface of ComPASS

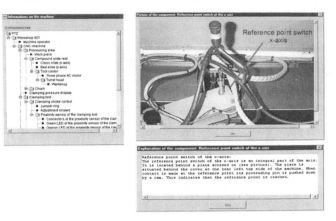

Figure 8.2: Screen shot consisting of: Component tree, photo of fault area, help-text (clockwise) (Marzi and John, 2001b)

The user-interface offers indications for control and correction measures as well as photos and explanations of components and assemblies. To achieve a larger comprehensibility, a tree of the machine components may be presented to the user. Machine operators are used to this form and therefore request it explicitly. Furthermore, information about the component's function and possible malfunctions is available. Figure 8.2 shows an example of a screen detailing the solutions simultaneously from different aspects. The component tree (upper left part) allows the user to find the source of the fault within the structure of the whole machine. The user can see, which other components are parts or neighbours of the source of the fault. For visual support to locate the fault at the actual machine, the relevant surroundings of the fault area are presented with pointers to the important parts (top right). In the text box at the bottom the repair process is detailed. The sequence of steps to be taken for the repair is described, so that the operator can follow the instructions given and performs the checks suggested.

Evaluation of ComPASS

The system was evaluated according to its usability and its effect on the change of the users' competence. The software-evaluation was executed by adapting forms from (Gediga and

Hamborg, 2000). In the context of the project, competence promotion was defined as the increase of qualification. It is perceived in the evolvement of the task-specific, methodical and actual preconditions for the workers' performance, which take effect during the task execution. In order to measure the effect on the users' competence, a knowledge test was developed including task executions (trouble-shooting) at the machine and a questionnaire for the evaluation of the fault situation. The degree of competence was measured with dependent variables such as different types of knowledge, diagnostic success, skills and stress (table 8.1). Each variable was determined by one or more questionnaires and practical tests (Krüger, 2000).

Table 8.1: Dependent variables and measuring them [based on (John and Timpe , 2000)].

Dependent variables	Measured by
Knowledge	Knowledge test (3 parts)
Problem solving capability	2 Fault-diagnosis routines in pre- and post-test
Self-assessment of competence	Competence questionnaire

The test persons were divided into two groups: the first using ComPASS for the test sequence, the second a special manual containing the same information. This control was necessary to ensure that the competence enhancement actually depended on ComPASS and not on the experience gained during the test itself. To measure by how much the competence increased, the complete test started with a pre-test without the use oft ComPASS / manual, continued with an intervention with the use of ComPASS / manual and ended with a post-test without the use of ComPASS / manual (table 8.2).

Table 8.2: Experimental design (Marzi and John, 2002a)

Pre-test	Exercise	Post-test
Instruction	Instruction	Instruction
Repetition of the introduction to the machine	Introduction to the MAS / manual	2 trouble-shooting routines
Studying the manual (5 min)	Studying the manual (5 min)	Questionnaire concerning the fault situation after each fault
2 trouble-shooting routines	4 trouble-shooting routines	Questionnaire concerning stress
Questionnaire concerning the fault situation after each fault	Questionnaire concerning the fault situation after each fault	Questionnaire to determine the competence of the test person
Questionnaire Concerning the stress	Questionnaire to MAS / manual	Knowledge test
	IsoMetrics for test group	

Each test consisted of several routines, where the test persons had to find and to remove faults of similar severity implemented by a machine operator. After the intervention every test-person working with ComPASS had to complete a questionnaire to evaluate the system. After completing the entire test the two groups showed different levels of competence, showing ComPASS contributed to an enhancement. The overall usability was rated as good. Some persons, who finished the tasks before the time elapsed, went on to try out the features and capabilities of the system. In order to measure long-term effects of the use of ComPASS, the system should be implemented in a company and used by the personnel over a time period of at least half a year. Further results are detailed in (Krüger, 2000).

Though the system raised the competence in the expected way, unexpected problems occurred concerning the desired usage, as not all features offered by the system were used. Based on the results (in detail see John and Timpe 2000) the user interface of the system was revised. Also, some changes to the underlying system had to be made. For the base system a programming language was chosen, which facilitated its quick implementation but lacked real-time capabilities. But these are paramount for ComPASS to be used in an SME. This made a re-implementation in a faster programming language necessary. Further, the application development engineer has to be provided with a tool to support input, change, and consistency checks of machine specific data. Thirdly, in the base system, the operator had to print out the results of the fault diagnosis (i.e. the recommendations for checking and repairing the machine), take the sheet to the machine and read while working. It was suggested, to provide the user with an acoustic means to be listened to during work. The new output medium had to be included while preserving the competence enhancement capabilities of ComPASS. For details on the speech output and its navigation structure see (Marzi and John, 2002b). The graphical user interface had to be changed due to the addition of more agents (protocol and tutorial agents), which required sequentialisation as well as partial hiding of information (Marzi *et al.*, 2002).

Integration of ComPASS

The main benefits of the integration of ComPASS into conventional diagnostic systems are the modularisation of machine specific knowledge and the capability of representing knowledge in different views. The agent-related information allows simple extensions and modifications of implemented system. The knowledge representation in different views shows the machine operator different solutions and leaves the final decision to him.

The system should be updated with newly acquired information (changes of functionality) from the machine manufacturer as well as the operators (maybe they found tricks or work-arounds), which enables the users to share knowledge. Workload and competence have to be adjusted accordingly by changing work and possibly responsibilities, work definitions (and maybe even payment schemes) have to be adjusted. Care has to be taken as to the rights to make changes to ComPASS. Its correctness and completeness have to be constantly ensured. In order for such a knowledge exchange to work, the operators involved have to be sure that this is not to their disadvantage. Otherwise, as stated earlier, technology cannot unfold its potential, not being culturally embedded in the organization.

CONCLUSION

As has been stated before, successful KM involves two aspects: people and technology. The latter has to be built and introduced in a way to ensure acceptance by the former. Different management systems show that only accepted procedures ensure the use of computer support. One detail-oriented approach is the system ComPASS. Long term tests of the actual competence promotion of ComPASS by actively involving the users in the decision making process still have to be carried out. As technology develops, organizational structures have to keep path to take advantage of support systems.

The advantages of IT solutions are that knowledge by few can be preserved to be given to novices. So, since there permanently is a shortage of specialized personnel, it is especially important to find a non-intrusive way of knowledge proliferation. IT gives the opportunity to disperse knowledge and distribute it. It helps various personnel keep the same standards of information and level of qualification. Learning "on the job" during job execution as suggested with ComPASS may increase competence in a fairly time-efficient way and in the context of a specific (fault)situation. This does not require a huge amount of abstract thinking. One main problem is user acceptance. It was discussed to introduce some kind of user model to save the state of knowledge and progress of a particular user in order to facilitate the interaction in ComPASS. This idea was discarded for various reasons, one being that the user might feel supervised and therefore would not be willing to use the system at all. The benefit of individualized interaction is outweighed by the wish for anonymity.

REFERENCES

Bechtolsheim, M. von (1993). *Agentensysteme. Verteiltes Problemlösen mit Expertensystemen.* Vieweg, Braunschweig.

Bergmann, B. (1994). Zur Lernförderung aus psychologischer Sicht. In: *Die Handlungsregulationstheorie* (B. Bergmann, and P. Richter, eds.). (in German), pp. 117-135. Hogrefe, Göttingen

Döhling-Wölm, J. (2000). *Wissensmanagement als Basis organisationalen Lernens.* (Knowledge Management as Basis for Organizational Learning). Unpublished report. (in German). Weiterbildungsstudium Arbeitswissenschaften; Univ. Hannover, Germany.

Eppler, M. (1999). *Wissensmanagement – Ein konzentrierter Überblick.* http://sowi.iwp.uni-linz.ac.at:8020/sww/wtrans/d2k/2k03eppler/Wissens_manag_eppler.html

Gediga, G. and K.-C. Hamborg (2000). IsoMetrics: Ein Verfahren zur Evaluation von Software nach ISO 9241/10. In: *Evaluationsforschung* (H. Holling and G. Gediga, eds.). Hogrefe, Göttingen.

John, P. and K. P. Timpe (2000). Kompetenzförderliche Multiagentensysteme zur Unterstützung der Diagnose an CNC-Werkzeugmaschinen. *ZWF Zeitschrift für den wirtschaftlichen Fabrikbetrieb*, **4**, 163-167.

Konradt, U. (1995). Strategies of failure diagnosis in computer-controlled manufacturing systems: empirical analysis and implications for the design of adaptive decision support systems. *International Journal of Human-Computer Studies*, **43**, 503-521.

Krüger, K. (2000). Entwicklung und Erprobung eines Instrumentariums zur Kompetenzmessung (Development and Trials of an Instrument to Measure Competence (in German). Unpublished master's thesis, Humboldt-Universität zu Berlin, Berlin, Germany.

Marzi, R. and K.-P. Timpe (2000). Competence promoting multi-agent systems in fault diagnosis of CNC-machine tools. In: *Advances in Networked Enterprises* (L. M. Camarinha-Matos, H. Afsarmanesh and H.-H. Erbe, eds.). CD-supplement. Kluwer, London.

Marzi, R. and P. John (2001a). Design and evaluation of a competence promoting decision support system. In: *Procs. IFAC-Conference on Human-Machine-Systems.* (Preprint), pp. 667-681, Kassel, Germany.

Marzi, R. and P. John (2001b): Supporting fault diagnosis through a multi-agent-architecture. *Journal of Mechanical Engineering*, **216**, Part B, 627-631.

Marzi, R. and P. John (2002a). Supporting maintenance personnel through a multi-agent system. In: *Procs. 6. IFAC Symposium on Cost Oriented Automation*, pp. 171-175, Berlin, Germany.

Marzi, R. and P. John (2002b). Akustische, mobile Fernunterstützung bei der Fehlerdiagnose an CNC-Werkzeugmaschinen. In *Elektronische Sprachsignalverarbeitung* 13. Konferenz, Studientexte zur Sprachkommunikation (R. Hoffmann, ed.), Vol. 24, pp. 104-111, Dresden, Germany.

Marzi, R., P. John and A. Busse, A. (2002). Kompetenzförderliche Multiagentensysteme - Diagnoseunterstützung für Maschinenoperateure aller Fertigkeitsstufen. *ZWF Zeitschrift für wirtschaftlichen Fabrikbetrieb*, **11**, 572-575.

Paivio, A. (1971). *Imagery and verbal processes*. Wiley, New York.

Wilke, H. (1998). *Systemisches Wissensmanagement*. UTB, Stuttgart.

9

QUALITY SYSTEMS: SUPPORT OR HINDRANCE FOR LEARNING

Björn Wahlström[1]

INTRODUCTION

Quality and quality systems have received increased attention in all industrial activities over the last ten years. The driving force for this attention can be found in the need for defining quality of certain products in objective terms and to ensure that a defined quality level can be reached on a continuing basis. Quality is in this context understood to be a set of attributes that characterise products, services and work processes on some more or less objective scales. The quality of a product could for example be defined in terms of dimensions of performance, features, conformance, reliability, durability, serviceability, perceived quality, and aesthetics. Quality of work processes is somewhat more abstract, but could for instance be characterised through people employed, resources and time spent, and methods and tools used in the work. It should be noted that quality always is placed in relation to a specific purpose, or in other words quality measures the fitness of a product, service or work process for its purpose.

Nuclear power plants use quality systems as an important part of the activities by which safety is managed. Quality is also an important concept in defining goals and assessing their achievement. Quality systems are required by nuclear regulatory bodies and the International Atomic Energy Agency (IAEA) has issued recommendations for how quality systems should

[1] Technical Research Centre of Finland, Espoo, Finland

be built and maintained (IAEA, 1996). Anecdotal evidence from the quality systems at nuclear power plants point to various problems in their implementation. A small study was therefore initiated within the Nordic Nuclear Safety Research (NKS, http://www.nks.org) to collect views on quality and quality systems. A total of 74 persons at the nuclear power plants in Finland and Sweden and at one research reactor in Norway were interviewed in the study.

The study brought many concrete suggestions for how quality systems should be built, adapted and integrated in the activities at the nuclear power plants. The demands set on hazardous industrial activities necessitate the implementation and use of a formal quality system, but there are many pitfalls, which should be avoided when such systems are built and used. The position of a formal quality system as a vehicle for organisational learning and knowledge management is discussed based on the results from the study. It is concluded that quality systems, which are built on participation and understanding, have a large potential of becoming efficient tools for organisational learning.

QUALITY IN THE NUCLEAR POWER INDUSTRY

The nuclear industry has become one of the most controversial industries in the world. The accidents at TMI and Chernobyl brought the hazards of nuclear power plants to the attention of the general public and they also made it clear for the industry that safety and high quality of all plants worldwide is a necessary precondition for continued operation. The nuclear industry has been a forerunner in the development of safety management activities as discussed in more detail below. The quality systems have an important role in these activities.

Requirements for nuclear power plant operation

The operation of nuclear power plants have characteristics similar to other industries that have a high accident potential such as air transportation and the chemical industry, but it is also different in some respects. Perhaps the most important technical difference is that nuclear power plants require a continued oversight even when they are shut down. The societal concern of risks connected to nuclear power, are much larger than risk estimates given by experts. The nuclear industry is global in the sense that bad performance in one plant has an impact everywhere due to decrease confidence and trust. All this implies that the burden of proof that continued operation is safe is far greater for nuclear power plants than for conventional installations.

The safety of the nuclear power plants builds on a number of safety principles of which the *defence in depth* is the most important. This principle implies that several independent barriers and safety systems are implemented to prevent the plants from entering a course of events, which may have disastrous consequences. In-depth safety analyses are used to ensure that all applicable safety criteria are fulfilled also during accident conditions. The *safety analysis report* is a comprehensive set of documents, which provides the basis for a plant owner to apply for a license to operate the plant and it also functions as a documented reference for safety management activities. The licensing conditions given by a nuclear regulator contain many provisions which should be attended continuously together with the requirement that formal reports should be issued whenever there have been deviations from the licensing conditions.

All these requirements taken together introduce the need for a formal system to regulate work activities at the nuclear power plants. The need for this system to be both comprehensive and well documented is further increased with the projected life-time of the plants, which even can be above sixty years, in combination with the large variety of skills which is needed for plant operation. In summary one could say that a formal system also brings the benefit of systematic and regular actions to correct for possible slow deterioration's, thus ensuring a covering collection of world-wide feedback of experience and standardised methods to ensure that experience is made easy to access.

An industry in change

The nuclear industry has gone through a considerable change since most of the present nuclear power plants were taken into operation. In the 1970ies the nuclear industry had a considerable political support in most countries. At that time it had no difficulties in attracting the brightest students from the best universities. This situation has changed radically today. Some countries have decided to phase out their nuclear capacity, although it may seem difficult to decrease the present share of nuclear power in the electricity supply (European Commission, 2000).

The deregulation of the electricity supply, which has taken place in many countries, has put a pressure on many nuclear operators to decrease their costs. The deregulation has also lead to a restructuring of the ownership of the electric utilities, which has been reflected in mergers and acquisitions. To adapt to the new situation several nuclear power plants in the world have initiated organisational changes, which among others include downsizing and outsourcing.

Some nuclear regulators have reacted to the changes in the industry with an expressed fear that important safety related knowledge could be in the danger of being eroded. Some regulators therefore require licensees to prepare a safety case before entering major organisational changes. An important part of such a safety case is an extensive survey of necessary skills and knowledge for running the nuclear power plant safely. Such surveys are further motivated by the fact that many nuclear plants worldwide are facing a generation change in their personnel.

Quality systems at nuclear power plants

A scientific approach to quality and quality control goes back to the early decades of the last century, when the building industry used collective experience to achieve high quality through repeatability in work processes (Slaton 2001). After the Second World War quality was introduced into Japanese management thinking by pioneers such as Deming. Later this development led to concepts such as Kaizen and Total Quality Management (TQM), which were used extensively in the Japanese car industry. The quality thinking was adopted gradually in the rest of the world through quality associations and the use of quality circles. Quality is now a well-established concept through the ISO 9000 series of standards and many companies today have certified their quality systems. Writing on quality can be grouped into three types, the prescriptive teaching of quality experts, quality certification and quality awards, and academic research (Ahire and Ravichandran 2001).

The nuclear industry joined the development of formal quality systems in the late 1960ies and early 1970ies. Initially the driving force was connected to the requirements for pressure vessels, but it was soon realised that the systems had a larger area of application. The concern for nuclear safety and the need for establishing systematic methods, by which a high repeatability in operations could be achieved, also contributed to this development. A working basis of quality assurance (QA) was first established in the nuclear field through early American legislation. Today national regulatory bodies require quality systems to be implemented at nuclear power plants, largely in line with recommendations given by the IAEA for how quality systems should be built and maintained.

The basic thinking in and philosophy of the quality systems in the conventional and nuclear industry is very much the same, but the historical development, paired with the special requirements for nuclear installations, has caused a difference in the details. A comparison of differences between the IAEA recommendations and the ISO 9000 standard can for example be found in IAEA (2000). In spite of national and international requirements and standardisation efforts, there are still variations in details of how different nuclear power plants have built their quality systems.

The perhaps most important trend in the quality systems at the nuclear power plants today is that they are integrated into the larger context of management systems. This development also includes the introduction and use of environmental monitoring systems. This development could be seen as a move towards a wider application of the TQM thinking in the nuclear industry. This is perhaps also to be expected, as TQM has been characterised to be among the most prominent operations improvement approaches of the twentieth century (Ahire and Ravichandran, 2001). In moving along this path it would however be important to understand why some companies have succeeded where others have failed in applying quality systems.

VIEWS ON QUALITY AND QUALITY SYSTEMS

In this section we report on a study concerning views on quality and quality systems. The data were collected through interviews with a total of 74 persons at the nuclear power plants in Barsebäck, Forsmark, Loviisa, Olkiluoto, Oskarshamn and Ringhals, and at the research reactor in Halden. The study was motivated by the importance quality systems have in a safe operation of nuclear installations. Anecdotal evidence of problems in the implementation of quality systems was also available as a motivator for the study.

The study

The study was initiated as a part of the SOS-1 project "Risk Assessment and Strategies for Safety" within the Nordic Nuclear Safety Research (NKS). The study was a follow up of a similar study investigating views on safety culture (Hammar *et al*, 2000). That study showed that aspects of safety culture are manifested in the quality systems and that safety culture on lower levels in the organisation is often associated with the quality system.

The aim of the study was to collect views and opinions concerning suitability and efficiency of the quality systems. As the systems at the various plants differed in certain respects, an opportunity was also seen in gaining understanding of the effects these differences. Another matter of interest was whether the rather elaborate quality systems gain adequate commitment on the part of all concerned. There have been some fears that such commitment might be lacking at the practical working level in the organisations, although high commitment to quality in a practical sense always has been present.

The persons interviewed were selected by a contact person at the sites to represent both developers and users of the quality system. The dates for the interviews were proposed by the contact persons in order to make a suitably diverse group of people available for the interviews. A categorisation of the persons interviewed showed that 49 represented users of the systems and 25 developers of them. Of the interviewed, 36 had a managerial position and 38 could be considered as having the position of an expert within their organisations.

Before the interviews took place participating organisations were asked to send descriptions of their quality systems to the interviewers. The interviews were carried out in the period 30.8-13.12.2000 and each interview took about one hour. In the interviews ten broad areas connected to quality were discussed (table 9.1). The selection of these areas was based on earlier experience and discussions with people from the nuclear field. They also provide a kind of logical sequence from the more general to the more specific, rounding

1. The quality concept
2. Quality systems
3. Topical quality related issues
4. Means to reach quality ends
5. Rules and procedures
6. Competency and training
7. Safety inspections and reviews
8. Process oriented activity control
9. Fostering quality thinking and commitment
10. Strategies and development needs for the future

Table 9.1: Areas discussed in the interviews.

off the interview with a discussion of the future. All interviews were taped and transcribed. The full report of the study has been issued by NKS (Hammar *et al*, 2001). Preliminary results from the study were presented and discussed at a seminar, which was held at the Ringhals nuclear power plant (Hammar and Wahlström, 2001). These discussions provided additional insights for the analysis of the results.

Reflections from the interviews

The quality concept. As was to be expected there was complete agreement that quality is essential in ensuring safety at the nuclear installations as well as for meeting other operational goals. Some people pointed at the need of accounting properly for all types of goals to be met in assessing quality of operation. Some expressed doubts, however, as to whether elaborate quality assurance concepts add significantly to people's generally rather obvious dedication to quality.

It appeared that quality was well understood in line with currently established definitions, i.e. in relation to requirements and expectations set for products and services. Many advanced the supplier's point of view relating quality largely to customer satisfaction in the broadest sense. Some took the view of the nuclear operator in emphasising the need of also gaining general approval of the enterprise by the society, together with confidence and goodwill. In this perspective the quality concept can be extended to apply, in general, to all activities involved in the operation of the nuclear power plants.

Some difference in the interpretation of quality could be seen depending on the role and function of the persons. People in managerial positions, for example, more often pointed to the need to define *sufficient* quality as compared with extravagant and unnecessarily expensive quality.

Quality systems are sometimes associated with bureaucracy and some people reported that they try to avoid the word quality. The bad ring of the word quality seems to be connected to the way early quality systems were introduced. Thus, a preference for speaking of, e.g., *operational control* instead of *quality of operation* can now be found.

Quality systems. The quality systems at the participating organisations were typically described in a top-down fashion starting with a quality policy, which is broken down into managerial directives and requirements to be applied at different organisational levels. The directives and requirements link further to detailed instructions and working procedures to be used in operation, maintenance, and technical support activities, etc.

The interviewed were generally quite satisfied with their own quality system, but they also indicated various needs for improvement. Examples were given of measures to verify that requirements made at certain organisational level indeed constitute a comprehensive response to directives and requirements by defining *quality demands* to be met by *quality responses*. A weakness commonly pointed at was that specific information could not always be found easily and quickly. Another weakness in the quality systems was that the links between the managerial requirements and the underlying instructions were not always seen clearly.

Good quality systems are associated with structure and understandability. A quality system has to be a living system, which means that it is updated regularly to reflect changes in organisation and practices. A quality system has also to be enforced through managerial example and actions. A good quality system is used in practice, as reflected by records of updates made in various parts from time to time.

A rather common view was expressed that the working staff generally is less familiar with the higher levels of quality system than with the lower level procedures and instructions and may therefore not see the quality system in its full context.

Topical quality related issues. Asked about what kind of quality issues currently is topical all indicated satisfaction that the operation at their plant is well under control and in compliance with the requirements for quality and safety. There are many activities continuously ongoing at the nuclear installations, which are related to quality and quality systems. Firstly according to the requirements of the quality systems themselves audits are conducted on a regular basis and remedial actions are taken in response to observations and deviations. Secondly various minor changes in the organisations bring in the need to update the quality systems and many such were under way at the installations visited. There are also quality issues raised in ongoing development programs which are concerned with documentation of rules and procedures, information technology, competence management, safety assessment practices, etc. Finally some of the visited organisations were involved in rather large modifications of their quality systems.

There was significant development activities under way in the organisations visited. These included quality audits, broadening the scope of individual audits to cover entire processes, working for a larger commitment of the top management, and involving the organisation as a whole in quality activities. They also included a transfer of emphasis to inspecting relevant activities instead of just collecting information in interviews and meetings. Finally there is an increased focus on the identification of root causes of observed deficiencies to identify efficient remedies.

Some of the nuclear power plants have on voluntary basis selected to comply with the standard ISO 14000 to minimise environmental impacts and to have this activity certified. Also plants which had not yet taken this route were preparing to take such steps in a near future. As a follow up of these activities many saw a benefit of a further integration of safety, quality and environmental issues into one management system.

Means to reach quality ends. Asked to indicate various means to reach quality ends, which require particular consideration to achieve continued improvements, it was largely pointed to documented procedures, instructions and handbooks. The utilisation of information technology was also seen as a way for improving access to the quality system.

Involvement and commitment on part of senior management in quality activities was generally thought to require further promotion. A high degree of involvement and participation from the whole organisation in all developments of the quality system was considered important in achieving commitment and efficient implementation of the quality system. Training in the quality system and more generally providing the reasons behind the system together with its bearing principles were seen as important.

The auditing process was emphasised by many as carrying further potentials for improvement in addition to being fundamental to quality. The audits are also considered valuable in spreading sound quality thinking and providing the reasons behind the system together with its bearing principles. The audits were also seen as a vehicle for insight and learning both in being audited and participating in the audit team. In some of the organisations the higher management took regular part in the audits and this had proven to be very useful.

Organisational structure was also pointed out as needing attention, e.g. in regard of managing various work processes involving several units in the line organisation. Another organisational issue pointed at, was the extent to which co-operative relations in the organisation should emulate those between sellers and buyers in order to emphasise mutual responsibilities. Recent experience has indicated that some caution should be observed in this respect.

Rules, procedures and instructions. Many of the interviewed were concerned about that their systems of procedures, instructions and handbooks had been allowed to grow too large. Reasons were given that deficiencies were previously often rectified by issuing supplementary instructions rather than by adjusting existing instructions. Unawareness of already existing, applicable documentation, due to lacking transparency of the documentation system also contributed and possibly also some craze for writing instructions. At many of the visited organisations work is now under way to reduce the number of instructions and at the same time improve the structure of the documentation system.

The point was often made that different types of instructions are needed of which some are intended to be followed step by step, where others are more for guidance and memory support. There were some remarks that detailed step by step instructions, while necessary in certain applications such as in the main control room, may not contribute to professional pride and commitment if used unnecessarily.

In a general comparison the operating instructions seem to be of high quality, while the maintenance instructions have somewhat uneven quality and the administrative instructions have the largest need to be improved.

It is evident that the instructions and procedures need to be updated regularly and that this work may become voluminous. To succeed this work has to be systematic and carried out to account comprehensively for all collected insight and experience. Most of the organisations have acquired computerised documentation systems to support the updating process. Many also indicated a belief that modern information technology has a potential to solve some of the problems as seen in present paper based systems.

Competency and training: All organisations visited conduct some kind of systematic competency inventory on a regular basis. In Sweden new regulatory ruling requires a systematic approach in performing such inventories. An extensive documentation concerning competency of operational staff and other staff involved in decisions or actions directly affecting the safe operation of the nuclear reactors was under way.

All organisations visited have individualised training systems in place, although some people gave examples of practical difficulties with the systems. Many asked for more training in quality and safety issues.

Despite the energy policy in Sweden, which implies a definite although yet not fixed time limit for operating the nuclear plants, there seems presently not to be too difficult to maintain the required level of competency. Many of the interviewed however, articulated fear that this might change on a medium term. In Finland some mentioned the possibility of a new nuclear power plant as one opportunity to attract new people to the field. On a longer term it is clear, however, that there will be considerable difficulties in maintaining competency in specialised nuclear professions.

Safety inspections and reviews: Safety authorities require plant owners to perform a large variety of safety inspections and reviews. These include reviews of safety related plant modifications, changes in operational procedures, event reports, etc. and also verification of operational readiness before start-up.

A safety inspection and review is conducted in the first place within the department responsible for the point at issue, while observing that staff which as been directly involved in work to be assessed will not take part in the assessment. As a rule, the inspection and review will be assessed independently by another party.

In Sweden the regulatory body requires that the plant owner makes independent safety assessments on its own to verify the quality of safety inspections and reviews made by the responsible department and to also make in-depth checks as deemed necessary. Such independent assessments are then organised through the plant quality department, which reports directly to the plant general management. In Finland the final safety review is made by the safety authority.

All interviewed expressed confidence in the inspections and reviews made at their plants, that the quality of particular assessments made as well as that all matters possibly involving risks are sufficiently well accounted for. The interviewed were convinced that independent inspections and reviews add to the quality of the activities as a whole. Some pointed out that there is a need for structuring inspections and reviews properly to utilise resources more efficiently. There were also remarks that it is necessary to ensure that the thoroughness of the initial safety assessment does not weaken because of undue reliance on assessments known to follow.

In Sweden the new regulation issued in 1998 has contributed significantly to clarifying regulatory requirements on safety assessments to be made by the plant owners. Several of the interviewed reported that the new regulation had been a valuable aid in structuring the safety assessment work.

Process oriented activity control: Industrial management can be regarded as a matter of controlling and coordinating work processes, like production, maintenance, procurement, development etc. Because there are many interacting work processes to be managed, process oriented activity control has been developed as a method to put a focus on the flow of work activities over organisational borders. A process view helps in detecting and correcting bottlenecks in handing over results from one work activity to another. The process view is sometimes seen as horizontal and complementary to the vertical view as provided by the line organisation. According to process oriented thinking it is a common practice to divide between core and supporting processes. Process orientation in the control of work activities has been introduced in many organisations, perhaps more outside than within the nuclear power industry. The process view is also supported by the ISO 9000 series of standards. A more detailed account of this methodology has been given by Rummler and Brache (1990).

Some of the visited organisations have introduced or are about to introduce process thinking in their quality systems, while others continue to use more traditional approaches in structuring their activities. There was a group of the interviewed that were not entirely familiar with the concept of process oriented activity control, but when the concept was explained, they had an intuitive feeling of its benefit. Most of the persons interviewed saw process orientation as a concept helping to structure activities in which many organisational units are involved. Requirements in regard of processes, when incorporated in the quality systems were seen to include requirements for each process to be clearly described and to assign responsibility for it to a *process owner*.

In spite of the benefits of process oriented activity control, there are still things to be resolved in trying to find a proper balance between the traditional line-organisation and a process oriented way to organise. Should both views be pursued in parallel or should one be selected before the other? Regulation implies the existence of a clear line of command and reporting which has to be merged with another structure, oriented along the processes. There were remarks to the effect that process orientation should not be understood to mean predisposition towards complete reorganisations along certain processes, but rather the attentiveness of the concept when possibilities for improvements are sought.

Process oriented activity control has been considered in some way or another in all the organisations visited. There were also references to cases where the process concept has played an important role in the development. For instance, there have been recent reorganisations at two of the visited plants to form a common maintenance department to units serve all production units at the site. In connection with these reorganisations thorough process analyses of the maintenance activities were performed.

Fostering quality thinking and commitment: In fostering quality thinking, many emphasised the need for applying a motivational approach instead of just considering the formal aspects of the quality system. That also implies that quality audits are seen as opportunities for improvements instead of a search for deficiencies. When this is achieved quality audits are capable of contributing to good quality thinking. Two of the organisations visited noted that they had positive experience of analysing deviations more thoroughly as symptoms of more general deficiencies. An active participation on part of the management as well as of peers from other departments or organisations in the audits are also due to increase their significance.

Commitment to quality requires, as several pointed out, knowledge and understanding of how it is connected to organisational goals. Commitment to quality can, for this reason, not be expected unless the quality system is well understood, e.g. in regard of definitions of authority, obligations and responsibilities. Fostering quality thinking is also connected to participation and efficient communication within the whole organisation. Many thought that promoting the process approach in work activities contributes to sound quality thinking and that it helps to give people a broad insight of their own roles in a larger context.

In the discussions several persons referred to the importance to understand organisational goals and how they are broken down to set the goals for organisational units. Some of the organisations visited have defined their major success factors and broken down them to provide a top down definition of a goal structure through organisational levels and even down to single individuals. This break down provides also a possibility to follow-up how these goals are achieved.

Several of the interviewed thought it would be important to create an understanding of the quality system and the ideas behind it, basically to tell what is right and why, and thereby to facilitate doing things in a right way. Information about the quality system should be given repeatedly and discussed actively, not only in training and education programs but also in management meetings as well as informal meetings arranged specifically for that purpose. Meeting days wholly assigned to related matters, like safety culture, were mentioned as highly valuable.

Favourable promotion on broad basis of quality thinking, specifically in regard or safety, was reported at two plants as a result of a long lasting campaign called STARK[2]. The campaign continues to be visibly pursued, by means of posters and gatherings receiving considerable support from all management.

Asked to judge on the general commitment to quality on part of their co-workers, all expressed full satisfaction. The high degree of commitment to safety was generally attributed to good traditions and culture developed in the nuclear field and not that much to the existence and content of the formal quality systems.

Strategies and development needs for the future: The largest future challenge is to maintain public trust and confidence in the safety of the plants while at the same time keeping them economically competitive. All this should take place in an environment where the safety requirements most likely would be increasing. Only then it would be possible to keep the

---

[2] Stanna, tänk, agera, reflektera, kommunicera (stop, think, act, reflect, communicate). Stark is the Swedish word for strong.

installations in operation for their remaining lifetime. Many referred in this connection to the deregulated electricity market and present difficulties in maintaining economic competitiveness with prospects of lower electricity prices. Many of the interviewed referred to safety as an absolute requirement, which is not possible to compromise in any situation. Others noted that a sustained safety could be reached only through the application of the principle of continuous improvements. Ensuring safety at constant high level requires, in practice, that there are continuing efforts expressly aimed at achieving further improvement of the safety. Several of the interviewed persons noted that if one would be satisfied with the present level, there is a risk for developing complacency.

Many of the interviewed in Sweden referred to the uncertainty for the future in regard of the political decision to close down the nuclear plants in a foreseeable future, well before the end of their economic life expectancy. The present strategy is to continue investment programs in safety and operational quality as if such a political decision would not come. This is applicable also for the reactor tentatively scheduled for closure in 2003.

Another challenge mentioned by many relates to the retirement of many of the present specialists in vital areas, within a decade. In Sweden plant owners are presently jointly surveying the situation for the nuclear plants in regard of the future availability of competency and other resources. A similar study was recently reported by a government working group in Finland (Ministry of Trade and Industry, 2000).

Observations from the study

The study confirmed that quality and quality systems have an important place in ensuring a safe operation of nuclear installations. There seems also to be a very good awareness and understanding of the demands, which are set on various activities to be acceptable. This awareness and understanding can be divided to address three different aspects of the systems, i.e. the existence of various threats, the actions, which can be implemented to meet them and the quality at which these actions are implemented.

Organising for quality: There are regulatory requirements on quality systems that they should be documented, reviewed and updated. These requirements have taken slightly different forms in Finland, Norway and Sweden. The organisations visited have selected different ways of structuring their quality systems. In spite of the differences the views on quality and how it can be ensured were rather similar. A view that quality assurance is a responsibility only of the

QA-department seems not to be valid. Instead there is a large agreement on that quality is a concern for everybody in the organisation. There is also a broad understanding of the need for formality in the quality systems. Still it seems that the quality systems in some of the organisations were better accepted than in other.'

Today the quality systems are expanded and merged with other systems to become integrated management systems providing a planned and documented account of all activities. In the management systems there is usually an easily traceable path from broad mission statements to the detailed instructions for tasks to be carried out. These systems then integrate all aspects of the enterprise such as business objectives, strategies and policies, rules, requirements, organisation and procedures. This goes along with a view that quality is built on awareness about goals, requirements and acceptable practices among all those taking part in the common mission of the organisation. This similarity between for instance quality, environmental monitoring and management in general has also been illustrated by Curcovic *et al* (2000).

The challenge in developing a good quality management system seems to be in finding a suitable structure, which makes it easy to navigate between principles, requirements and solutions. Another challenge is to break down general goals and requirements to give practical guidance for all work activities. A successful combination of all these requirements in the management system with a due account of both the line organisation and process-oriented activities will require some innovative thinking.

Goal definition and follow up. The formulation, prioritisation and follow up of goals are important parts of activities aiming at quality. Definition of goals and follow up how they are achieved, typically go through cycles of strategic and yearly planning. Today there is a tendency to break down company goals in several hierarchical steps even down to the level of individual persons. This is often done rather formally in the framework of the so-called balanced scorecards, which were introduced by Kaplan and Norton (1996). According to the concept the overriding goals or success factors for an organisation are defined in a systematic way and to further broken down into sub-goals. One example of the high level indicators on the balanced score card used in one of the organisations was *production, safety, economy* and *public confidence.*

The balanced scorecard concept was seen to enhance a participation in goal formulation throughout the whole organisation. Following up to what extent goals have been achieved take a reversed process starting from individual performance appraisals and ending with an assessment of the indicators on the balanced scorecard. Many of the interviewed saw the balanced score card as a functional approach, in emphasising a selection of fundamental areas in which goals have to be reached for ensuring survival of the organisation.

Some of the interviewed pointed to the need for arriving at a proper balance between the sometimes conflicting goals as set on the nuclear installations. Safety culture has for some time been seen as a goal in itself and has been strongly promoted by international organisations and national safety authorities. The strength of safety culture is however difficult to assess and quantify.

Some remarks on the often discussed conflict between safety and economy in nuclear operation were made and it was argued that such conflicts disappears in a longer time frame, because economy can only be ensured if safety is demonstrated consistently. There were, however, also remarks that the increasing economic pressure due to the deregulation of the electricity markets may accentuate the conflict. This may indeed be a danger if it is not forcefully resisted, because even safety improvements tend now to be postponed if not considered to be absolutely necessary.

Inspections, audits and reviews for quality: Inspections, audits and reviews are important activities with which a continued quality in the work is maintained. All components and equipment that are procured are inspected to ensure that they fulfil their requirements. Inspections are made when components and equipment are installed at the plants to ensure that everything has been correctly done. For more complicated inspections an inspection plan is usually prepared in beforehand.

Regular audits of all activities or parts of the organisation are made according to stipulations in the quality systems. In an audit, a small team goes through the activity or the work of an organisational unit in a large degree of detail. Audits usually aim at detecting discrepancies between defined and actual ways of carrying out work. One common observation from audits is that the same observations tend to repeat. This point to underlying causes that have not been brought into the open. Several persons thought that observations from the audits should go through the same formal analysis as incidents.

The regulatory bodies are carrying out regular inspections, audits and reviews. Plant modifications with a safety impact are usually inspected by regulatory representatives. Quality and safety audits are done every year on slightly varying activities. A larger safety review consisting of several different areas is typically required with a ten-year interval and this includes both self-assessments by the nuclear utility and reviews by the regulatory body as described by IAEA (1994).

Audits are sometimes seen as to be more concentrating of doing things right than doing the right things. Activities should therefore also be reviewed in a more comprehensive way. Various review methods have been developed, and are in regular use by the nuclear power plants. International Atomic Energy Agency (IAEA) and World Association of Nuclear Operators (WANO) are offering peer reviews where a group of outside experts during a period of two to three weeks make a comprehensive review of many different safety related activities.

Use of information technology: It is evident that modern information technology offers several opportunities in making the quality systems easier to use and update. The accessibility of the quality system can be improved using computerised information systems with hyperlinks between different parts of the system. The transfer of procedures and documentation to a suitable computerised platform has a large potential as compared with traditional paper based systems.

All organisations visited were using e-mail and some kind of Intranet and people expressed a large satisfaction with these systems. Some pointed to the possibility of information overload when it has become too easy to send everything to everybody. There is on-going activities at all organisations visited which aim at providing information about their quality system through their Intranets. In an assessment of a future development it is possible that an increased use of information system may introduce a gradual shift to a situation where people are not passively fed with information, but are supposed to actively search for what they need.

Some of the interviewed indicated needs for improved handling a variety of administrative information, such as meeting protocols, decisions on actions to be taken, deviations identified in audits, etc. If that information can be accessed a compiled easily it would be easier to manage the follow-up activities to ensure that important issues are dealt with according to plans.

One drawback with modern information technology is that the systems tend to become obsolete very quickly. This introduces a burden in the training of people. The systems are also expensive and there are several examples of overruns both in costs and time when new information systems have been installed.

Exchange of experience: The study demonstrates large similarities both in solutions and in ongoing activities at all the visited organisations. Still there are relatively few direct contacts between the organisations. One could for example have expected joint seminars for exchanging information on ongoing activities and exchange of auditors for the quality audits. Active contacts were more directed towards technical details for example within working groups aiming at establishing common interpretations of certain standards.

The contacts between the nuclear power plants for an exchange of information seem to have been decreasing over the last ten years. One likely reason is linked to the increased competition brought by the deregulation of the electricity market. The deregulation has also had an effect of forcing nuclear power plants to reduce their costs, which may have had an influence on activities that are considered less important.

In the organisations visited there is no confidential reporting system like those in use for instance at many airlines. According to regulatory requirements all incidents above a certain level of severity are reported to the regulatory body. These incidents are analysed using formal methods for identifying root causes. There are arguments that confidential reporting could stimulate a more covering reporting, but the present practices seem to be able to capture the most important lessons due to two reasons. Firstly the reporting atmosphere is open to encourage people to report mistakes they have made. Secondly most incidents, which contain a safety challenge, tend to be complicated and involve several actors making it less likely that they will not be reported. Such incidents also make a formal analysis necessary for all the lessons to be learned. Some of the plants visited have a system for reporting minor problems, but they are more to be seen as a way to stimulate suggestions for improvements from the shop floor

Anchoring the quality system in the organisation: The persons who were responsible for quality activities mentioned as their largest challenge to get the quality system anchored in the organisation. Many of these persons saw their job as service activity in the organisation to promote quality thinking and good practices in all activities. It is evident that a system, which only is seen as a binder collecting dust on a shelf, cannot have its desired influence on the quality of work. There seems to have been earlier excesses in the quality assurance activities which have given the word quality a bad ring, but these problems have evidently been cleared away today.

Most people viewed the attitudes of the senior management as crucial in getting an organisational commitment to quality. Some of the persons interviewed who had a managerial position even expressed some astonishment with the small interest for quality activities that some senior mangers showed. A visible participation in quality activities could be seen in a larger attention to issues concerned with quality in that organisation.

A broad participation in the development of the quality system was seen to carry a potential of larger commitment to the system. Someone has to take the lead when a new system is developed and introduced, but it is important that everyone has a saying and that the system is systematically anchored. A novel view on quality system used in some of the organisations is that management formulates requirements for the organisational units and they in turn describe their responses to these quality requirements.

In the organisations visited there was an ongoing discussion on the application of a process thinking as a mean to identify quality problems in the interfaces between organisational units. Many of the interviewed saw this discussion as a way to anchor quality thinking in a broader context and to support a communication of understanding of how quality is built in consecutive work processes. The process thinking was seen as a method for making work activities smoother when organisational borders are bridged, but none of the persons interviewed saw this as being a realistic substitute for traditional line organisation.

Regulatory oversight: In the study no explicit question was asked on regulatory oversight. In spite of that, many of the interviewed made several references to regulatory activities. Differences in national regulation as well as the difference between commercial and research reactors were clearly reflected in the interviews. The structure and organisation of the regulatory activities in Finland and Sweden concerning the commercial nuclear power plants is also somewhat different as described by Wahlström *et al* (1996).

Regulatory activities have a large influence on quality-related activities at the nuclear installations. This is certainly also the intent, but there was a clear view expressed that the authorities should avoid having a too large influence on the alternatives selected for structuring work processes. This view is well in line with the bearing principle that the plant owners should have an undivided responsibility for safety. This implies that it is important to maintain a distance in the regulatory involvement. Additionally it is important that the regulatory attention is balanced in that respect that issues are given the weight they deserve by their safety importance. The study also supports the result of other studies (Sinkkonen, 1998) and Wahlström and Sairanen 2001), that requirements which are considered legitimate get a larger acceptance at the nuclear power plants.

In spite of the need for a certain distance between the regulator and the operator of a nuclear installation it is necessary to maintain open and trustful dialogue in all interactions. Such a dialogue seems to be facilitated by a mutual understanding of roles and practices. An important part in the processes for creating a pertinent regulatory oversight is also connected to activities of the regulatory body for developing its own competency and practices as reported by Reiman and Norros (2001).

IMPLICATIONS ON LEARNING AND KNOWLEDGE SHARING

Nuclear installations use formal quality systems to ensure accuracy and repeatability in all important activities. At best this quality system institutionalises organisational learning in a formal system into which the principle of continuous improvements has been built. At worst the formal quality system could be a bureaucratic burden which stifles individual initiative and thereby becomes a hindrance for learning. If the formal quality system is constructed and implemented with due consideration of the people who are going to use them there should not be any obstacles for the systems to be efficient vehicles for organisational learning and knowledge sharing.

Reasoning about quality and safety

The concept of quality is important for all work processes. The word quality embodies a set of attributes or properties to some object i.e. a product, service or work process, which can be measured on a scale to identify a region of acceptability. This region of acceptability is connected to a customer defined in a broad sense. Sometimes quality is seen in relation to expressed, tacit and even unconscious needs of this customer. Quality of a product can often be expressed in quite concrete terms, but quality of work processes tend to be more abstract. A large part of the quality requirements set on a certain product, service or work process are seldom fully expressed, which means that transactions between two or more people often need a clarification of tacit assumptions in the characterisation of quality. This actually implies that quality in a more formal sense cannot be defined without several consecutive loops of externalisation, combination, internalisation and socialisation as described by Nonaka and Takeuchi (1995).

Causation is another concept, which is important in a consideration of quality. One has to be able to reason about causes of good or bad quality. Such causes can be imbedded in work processes and the materials used, they could depend on knowledge and skills of the people doing the work, and could also depend on processes of chance beyond the control of anybody. The quality systems build on the assumption that quality can be controlled by influencing the work processes. An unsuitable work process together with negligent inspection may result in inferior quality with further consequences on safety and/or economy. The quality systems therefore include review activities aiming at identify reasons for inferior quality and possibilities to introduce remedies.

The concern for safety was the driving force behind the application of quality systems in the nuclear industry. It is easy to understand that inferior quality can cause safety problems, but it is also important to see the quality systems as providers of efficiency. The relationship between safety and quality should be mapped and considered to define the specific quality requirements by which safety and economic performance can be reached. The projected lifetime of a nuclear power plant also puts a demand on a comprehensive documentation of quality requirements and the work processes by which this quality can be reached. In this light the formal quality systems that are used at nuclear installations can actually be seen as a construct driven by safety requirements.

A functional view on quality systems

Quality systems can be seen as a mean to reach certain ends. The quality systems therefore have a well-defined function. On a basic level one could say that they contain descriptions of the agreed quality that has to be reached, methods for reaching this quality and the processes by which the quality system is maintained. From a purely functional point of view it does not matter if the quality system is tacit and embedded in the work practices or if it is explicitly described in a set of documents. From a practical point however, there is a need to have it formalised in one way or another to make it accessible and reviewable.

When a quality system is created there are several dimensions of freedom. Should for instance the system cover only the quality of products that are manufactured or should it also regulate work processes of the organisation? If a formal management system is used it is natural to include the quality considerations in this system. A second question is if the same quality system should apply to all parts of an organisation or is it acceptable that different organisational units have their own quality systems? Even if several different systems are used it seems however practical to tie them together in some high-level management documents. If different quality systems are used in the same organisations it is also practical to require a reasonable similarity between the systems.

Given the freedom in building a new quality system, one could ask what the characteristics of a good quality system would be. The process of building up a quality system gives many insights for the people involved in that work, but these insights have to be transferred to the whole organisation. One important characteristic is that the quality system must be easy to understand and apply. Understanding can be facilitated if the quality system is well structured and based on an explicitly expressed philosophy. A quality system can be brought in with force and excessive training, but a system that people can accept and apply by heart is always more efficient. Finally a quality system should have the full support of the management and it should be updated at regular intervals.

Perhaps the most important function of a formal quality system is to set an explicit requirement on systematic audits and reviews at regular intervals. In this function the quality system acts in a similar way as outside triggers in initiating a search for ways to improve. The trigger built into the quality systems is also enforced by the regulatory agency, which for example will ask questions about audits and their results. The quality system can in this way be viewed as a tool for continuous performance measurements and internal benchmarking.

The quality system as vehicle for learning

The quality system typically encompasses practices for activity planning, documentation and performance evaluation. Quality system can therefore provide a vehicle for learning if this purpose is considered and adapted to. This possibility has to be put in relation with the fear that a formal system may stifle innovation. From the point of organisational learning the question of whether or not to have a formal quality system, is reduced to a question of a preference between formal and ad hoc procedures for management. It is clear that a formal system never can make ad hoc procedures unnecessary, because no system can be designed to foresee all future needs. On the other hand a reliance on only ad hoc practices becomes inefficient if there are no systematic procedures to capture experience and make it operational. The need for formal quality systems therefore depends on the degree of uncertainty in the organisational environment.

Formal quality systems assume audits at regular intervals and they therefore facilitate systematic handling of operational experience. Audits and systematic analysis of events generate recommendations for improvements. These recommendations are transformed to specific actions to be undertaken and the actions are followed up. A formal quality system enforces reconsideration of old strategies and practices at regular intervals and it could therefore be seen as one example of the learning agent, which is discussed in the Chapter 2.

Audits and the analysis of events are typically done in small groups, which combine different specialisation and skills. This practice has the benefit of supporting the emergence of a shared language and understanding between different professions within the organisation. The quality auditors can also act as agents for the transfer of good practices, where both the audited part of the organisation and the auditors are exposed to different sets of assumptions and practices. Similarly participants in peer reviews have many opportunities to transfer good practices between different nuclear installations. To be efficient in this respect it may be necessary to give the auditors and reviewers training in this aspect.

The quality systems, and more generally, the practices, which are used in the nuclear industry for utilising operational experience, have large potentials to support organisational learning provided that the obstacles can be overcome. Perhaps most important is the attitudes people have towards the quality systems. If they are viewed as bureaucratic, difficult and not worth the effort, it is not likely that they can support learning. A second obstacle is connected to the problem of finding suitable concepts and models with which the bulk of operational experience can be coded to make sense and be accessible in a formal system. A third challenge is connected to the difficulty of translating experience coming from one nuclear power plant to concrete actions in a completely different setting.

Finding a balance between various extremes

The management at a nuclear power plant and in organisations in general has to find solutions, which satisfy goals and requirements. Sometimes this implies finding a balance between two extremes that appear conflicting. In a discussion of quality, such a balance has for instance to be found between quality and price or perhaps more correctly between defined quality requirements and the price that has to be paid to get it. This balance involves many considerations, because it may for example be advantageous to standardise to a higher quality, which then can mean that unnecessary high quality is used at some places. Relationships between quality and cost can be quite complex and express dynamics over time as Oliva and Sterman (2001) point out.

The quality systems give guidance for various work processes and they will therefore at least implicitly define how much effort that is spent reaching the required quality in the work process itself as compared with efforts spent afterwards on inspecting and correcting the work. The crucial factor here is the variability in the output as introduced by the work process, where a low variability can eliminate the need for extensive checks. According to commonly accepted views it is not possible to achieve quality just by inspections and reviews. This also means that quality deficiencies have to be caught as early as possible in the work processes.

Building up and maintaining a quality system involves finding balances between several extremes. The perhaps most important is a selection of the degree of formality to be used in the quality system. In a very formal system there are clear rules for how to act in various situations, but such a system may become very difficult to work with. A quality system should also be reasonably documented, but too many documents in the system may render specific documents difficult to find and access. The extent of a similarity in structure and content of the quality systems used in different parts of an organisation is also an important balance. A similarity can enhance communication and understanding, but it may be unpractical taking into account the specific needs in different parts of the organisation. All these questions may stir up emotions when two organisations are brought together in a merger and there are two quality systems that are built on different principles.

Quality systems can be seen as a response to the need for conservatism and stability of the nuclear industry. All solutions adopted for various purposes should in principle be proven to avoid the danger of later problems with immature solutions. This will to some extent slow the process of learning, but it will on the other hand prevent the organisation from jumping on all new management whims. Quality systems can support evolutionary developments, but will most likely tend to stifle development in cases where more revolutionary changes are needed.

Implications for organisation and management

The quality systems tend today to be integrated into the management systems. This is practical because quality is actually one of a variety of concerns, which can be handled with similar methods. There are a variety of standards and guidelines, which give excellent advice on how to set up and maintain the systems. The difficulty however, is that the guiding documents tend to be thick and somewhat difficult to access. An illustration of the underlying principles may help in this regard. A second observation is that the standards and guides do not usually give much guidance in how to make the systems user friendly and accepted.

The increasing complexity of industrial systems has brought an insight that a confidence in quality can be obtained only by a simultaneous consideration of both the product itself and the work processes that have generated this product. To verify that a product fulfils a certain quality level it would therefore be necessary to get an access to a large amount of information from the initial phases of design and use of the product. Unfortunately there are many products on the market, for which this kind of information can be very difficult to get.

When a formal quality system is developed it is necessary to have a person or a group of persons to whom the task of creating and documenting it is given. To ensure a successful result it is however necessary to have enough consultations with the persons who are going to use it. Some of the organisations participating in the study even expressed the view that the persons supposed to use instructions should be formally involved in creating them.

The quality system can be seen as a normative framework, which defines a formal organisation as opposed to actual or informal ways of organising work activities. There is anecdotal evidence that the formal and informal organisations may start to diverge and there may be various reasons for this. One possible reason is that the basic assumptions and philosophy of the quality system is not adapted for its intended purpose in the organisation. Another possible reason is that the quality system has not been anchored in actual practices on the shop floor. A third reason may be the absence of sound human factors principles in the creation of the quality system.

A vision for the future

Quality system will also in the future be one important part of the safety management activities at nuclear installations. Experience from the use of the quality systems has been accumulated in the systems themselves and also in various standards and guidelines for their design and operation. This development over the years as promoted by several organisations can be seen as a kind of organisational learning on a global level. In spite of the improvements made so far, there seems however still to be room for further improvements. This is perhaps a reflection of the general strategy of continuous improvement. Without this strive in the organisation there is the danger of complacency.

Presently there are considerable differences in the structure and details of the management and quality systems used at the nuclear installations. There are also differences between the management and quality systems used in the nuclear industry and in other safety-related industries. The application of the TQM concept in a wide sense has a potential to narrow these differences to make the views on the quality systems to converge. On a medium term it may be expected that the quality systems will become more similar. Such a development will however rely on a collection and analysis of experience from various quality systems to evaluate the characteristics, which make them fit for their purpose.

Quality systems are more geared to learning from failure than success, but due to the way audits and reviews are organised they have the potential also to transfer good practices. In the future this part may need strengthening. Another reason to stress also learning from success is that disappearing failures it would remove events at which learning takes place. To counteract this other learning mechanism may be necessary. One possibility may be to have some kind of institutionalised imagination to envisage how causal factors behind an incident somewhere else could be transfer to a completely different environment as March *et al* (1991) are discussing. This would need formal analysis and suitable methods and theories to make a believable shift of an event from one cultural setting to another.

CONCLUSIONS

Man is adapted to learn by trial and error. There are many systems in use today where this approach to learning cannot be tolerated. Experience from high reliability organisations has brought many insights by which activities can be made very safe (Rochlin, 1999). Experience however also point to mechanisms, which may introduce hidden deficiencies in the safety management activities as illustrated by Baumont *et al* (2000). The challenge is to detect and correct such deficiencies before an incident has made them obvious. The quality systems have an important function in this endeavour.

Nuclear installations are in many respects similar to other installations where a high safety is required, but there are also important differences. The quality systems that are used in the nuclear industry provide a reflection of these requirements. The nuclear industry is presently in the middle of a change process, which will be reflected also in the quality systems in use. Changes bring opportunities for learning and renewal, but they also carry an increased vulnerability that something important is forgotten. Opponents to nuclear power have claimed that it is impossible for any organisation to live up to a demand of zero errors and that accidents therefore are inevitable. The experience collected so far from the use of quality systems and from safety management more generally, has demonstrated that there are no obstacles in ensuring a continued safety of the nuclear power plants in the world.

The study gave many useful insights in the position the quality systems have as a part of the safety activities at nuclear installations. Given the historical background of the organisations it is often easy to understand solutions selected and positions held. Sometimes however, this historical ballast gives a feeling that it would be easier to start with a clean table, but the need to consider accumulated knowledge seldom makes this possible. The interviews brought many concrete suggestions for how quality systems could be improved and integrated in the safety activities at the nuclear power plants. There is clearly a need for a broad participation in the creating and maintaining of the quality systems. It is also important that the principles of the systems are understood and accepted. If that is not achieved they have no function and cannot consequently fulfil their purpose of controlling work activities. Properly used the quality systems have a potential of becoming good tools for organisational learning.

Finally there seems to be a need for a coordinated effort in approaching some of the important issues connected to quality systems used in the nuclear industry of today. These include, but are not restricted to; the structure of the quality system, its implementation and the ways to get it anchored into the organisation. The quality system should provide support for their users in making sense of the requirements that are placed on various activities and of the methods selected to meet those requirements. Quality systems are often viewed as rather technical, but they certainly have to do with people and how they communicate. Many managers at the nuclear power plants agree on that there is a need for development activities, but it seems difficult to find a natural body, which could approach this challenge with the correct blend of both theoretical and practical skills.

Acknowledgement

Support in writing this chapter from the Nordic Safety Research (NKS) within the project SOS-1 "Risk Assessment and Strategies for Safety is gratefully acknowledged. This chapter builds on many discussions, comments and insights from Lennart Hammar, Bengt Lidh and Teemu Reiman who participated in the interviews and in the writing of the report from the project. Olle Andersson from the Forsmark nuclear power plant gave many practical suggestions for improving the chapter. The most important contribution however, came from the 74 persons from the nuclear power plants in Barsebäck, Forsmark, Loviisa, Olkiluoto,

Oskarshamn and Ringhals, and at the research reactor in Halden, who shared their knowledge and experience on quality and quality systems with the interviewers.

REFERENCES

Ahire, S. L. and T. Ravichandran (2001). An innovation diffusion model of TQM implementation, *IEEE Trans. Eng. Manag.*, **48**, 445-464.

Baumont, G., B. Wahlström, R. Solá, J. Williams, A. Frischknecht, B. Wilpert and C. Rollenhagen (2000). Organisational Factors; their definition and influence on nuclear safety, VTT Research Notes 2067, Technical Research Centre of Finland, Espoo, Finland, ISBN 951-38-5770-0.

Curcovic, S., S. A. Melnyk, R. B. Handfield and R. Calantone (2000). Investigating the linkage between total quality management and environmentally responsible manufacturing, *IEEE Trans. on Eng. Manag.*, **47**, 444-464.

European Commission (2000). Green Paper: Towards a European strategy for the security of energy supply, COM(2000) 769, 29.11.2000.

Hammar, L. and B. Wahlström (2001). NKS/SOS-1 Seminar on quality assurance, report from a seminar in Ringhals 16-17 January 2001 (in Swedish).

Hammar, L., B. Wahlström and J. Kettunen (2000). Views on safety culture at Swedish and Finnish nuclear power plants (in Swedish, Syn på säkerhetskultur vid svenska och finska kärnkraftverk), NKS-14, ISBN 87-7893-064-2.

Hammar, L., B. Wahlström, B. Lidh and T. Reiman (2001). Views on quality assurance at Finnish and Swedish nuclear power plants and the Halden reactor (in Swedish, Syn på kvalitetssäkring vid finska och svenska kärnkraftverk samt vid Haldenreaktorn), Nordic nuclear safety research, NKS-38, June 2001, ISBN 87-7893-090-1.

IAEA (1994). Periodic safety review of operational nuclear power plant, Safety Series No. 50-SG-O12, International Atomic Energy Agency, Vienna.

IAEA (1996). Quality assurance for safety in nuclear power plants and other nuclear installations, SS No. 50-C/SG-Q, International Atomic Energy Agency, Vienna.

IAEA (2000). Quality assurance standards: comparison between IAEA 50-C/SG-Q and ISO 9001:1994, International Atomic Energy Agency, IAEA-TECDOC-1182, November.

Kaplan, R. S., D. P. Norton (1996). *Translating strategy into action: The balanced score card*, Harvard Business School Press, Boston, Mass.

March, J. G., L. S. Sproull and M. Tamus (1991). Learning from samples of one or fewer, *Org. Sci.*, **2**, 1-13.

Ministry of Trade and Industry (2000). Know-how working group, Chairman Carita Putkonen, Secretary Timo Haapalehto Measures to retain expertise in the nuclear energy sector, Working Group Report, Ministry of Trade and Industry, 10/2000, Helsinki, Finland.

Nonaka, I., H. Takeuchi (1995). *The knowledge creating company*, Oxford University Press, New York.

Oliva, R. and J. D. Sterman (2001). Cutting corners and working overtime: Quality erosion in the service industry, *Manag. Sci.*, **47**, 894-914.

Reiman, T. and L. Norros (2001). Regulatory culture: Balancing the different demands of the regulatory practice in nuclear industry.

Rochlin, G. I. (1999). The social construction of safety, in *Nuclear safety: A human factors perspective* (J. Misumi, B. Wilpert and R. Miller, eds.). London: Taylor & Francis.

Rummler, G, A. and A. P. Brache (1990). *Improve performance: How to manage the white space of organisation chart, second edition*, Jossey-Bass publishers, San Francisco, 227p.

Sinkkonen S. (1998). Utility views on nuclear regulatory oversight (in Finnish, Ydin-turvallisuusvalvonta Imatran Voima Oy:n ja Teollisuuden Voima Oy:n edustajien silmin), unpublished thesis, Helsinki University.

Slaton, A. (2001). "As near as practicable" Precision, ambiguity, and the social features of industrial quality control, *Technology and Culture*, **41**, 51-80.

Wahlström, B., R. Nyman and T. Reiman (1996). A comparison between regulatory activities for nuclear safety in Finland and Sweden (in Swedish, En jämförelse mellan myndighetsarbetet inom kärnsäkerheten i Finland och Sverige), NKS/RAK-1(96)R7, Oktober.

Wahlström, B. and R. Sairanen (2001). Views on the Finnish nuclear regulatory guides, available at http://www.stuk.fi/english/convention/yvl-review.html.

10

CIRAS – A METHOD TO PROMOTE KNOWLEDGE SHARING IN THE UK RAILWAY INDUSTRY

Linda Wright[1]

INTRODUCTION

This chapter presents an overview of the way knowledge of safety concerns is gained and shared by CIRAS, the Confidential Incident Reporting and Analysis System. CIRAS was originally developed and implemented by the University of Strathclyde for ScotRail Railways in 1996. Following the success of the system, it was expanded in June 2000 to include the safety critical and safety related staff of all railway related companies who are part of the UK Railway Group – a potential reporting pool of some 70,000 staff. After the first 18 months of nationalisation a total of 1,020 reports were received.[2]

The success of any safety management system is dependent not only upon the reporting of concerns and errors by staff, but also upon the willingness of management to accept and act upon the information. The CIRAS system is presented briefly, and the strengths and weaknesses of the system in terms of organisational learning are discussed.

---

[1] Formerly: CIRAS, University of Strathclyde, Glasgow, Scotland
Now: ProRail, Department of Safety and Environment, Utrecht, The Netherlands
[2] The national CIRAS system described in this chapter was operational from June 2000 for 18 months. Since 2002 there have been significant changes to the Core Data Centre organisation and the human factors system described here is no longer operational.

BACKGROUND TO THE CIRAS SYSTEM

CIRAS is the Confidential Incident Reporting and Analysis System currently being used in order to identify and deal with human factors problems on the railways in the UK. CIRAS was initially a response to the discovery that human factors (including human error and latent failures) contribute substantially to incidents, situations and near misses on the railways. An earlier background report by Vosper Thornycroft (Heybroek, 1995) pointed out the role of human factors in the rail industry, and the importance of these has also been highlighted in other industries (e.g. the off-shore oil industry; nuclear industry). Furthermore, existing official reporting procedures are often associated with disciplinary action, and this distorts the nature of reports received. This is particularly true in the UK rail industry where, historically, relationships between workforce and management have sometimes been characterised by mutual mistrust and animosity, rather than co-operation. The result is a tendency for reports to become focused on technical failures and chance happenings (the reports tend to be strategic, defensive and "external") with the human element being virtually absent (Van Vuuren, 1998). In some instances, it may even be the case that a near miss or incident with no obvious consequences will be deliberately concealed (i.e. the person concerned feels lucky to "have got away with it this time") due to the perceived disciplinary implications, rather than being seen as something from which others could usefully learn. The system was therefore introduced to collect and learn from railway near misses in order to compare them with and prevent more serious incidents (Wright, 2000), which would otherwise not be detected by the company.

Following the success of the system within ScotRail, other railway companies joined the scheme voluntarily. This culminated in the system becoming national and mandatory in June 2000, after an industry-wide conference took place and a number of highly serious railway accidents occurred. In addition to Railway Safety Ltd (now called Rail Safety and Standards Board), each company provides financial support for the system depending on the number of staff that could potentially use the system. In order to expedite the expansion of the system three Regional bases were established across the UK (University of Strathclyde, Glasgow; WS Atkins, Warrington; and DERA, Farnborough). These Regional bases fulfil the function of collecting reports, passing information to managers and publishing a regional Journal. In addition they also pass the collected information to the Core Data Centre where it is further analysed in terms of a human factors model and root causes. The Core produces a comprehensive six monthly report and also responds to requests for specific analyses from the individual companies (provided commercial information about a direct competitor is not requested). For a full explanation of the way the system operates see Davies, Wright, Courtney and Reid (2000).

THE AIMS OF CIRAS

The aim of CIRAS is firstly to collect reports relating to safety concerns and failures from safety-critical and other personnel employed in the rail industry. Secondly, CIRAS aims to contribute to the prevention of accidents and incidents by the sharing of knowledge of causal factors with individual companies. Thirdly, CIRAS aims to raise awareness of risk factors by reporting back to the companies and via the Journal. Fourthly, CIRAS aims to contribute to the sharing of such knowledge across the whole of the railway industry via the compilation of a single industry-wide database. The data and the associated data-analyses are made available to the whole industry - whilst the identity of everyone who sends in reports is protected and treated with the utmost confidentiality. The result is a body of information on which the industry can take action, or on which individual Train Operating Companies (TOCs) or infrastructure companies may wish to act. Within an industry consisting of some 70 private companies the need for a centralised human-factors resource of this type, guaranteeing confidentiality and independent of company interests, from which common as well as specific lessons can be learned, is clearly indicated.

Learning from Mistakes

Firstly, CIRAS aims to stimulate staff to share their knowledge of railway situations by making safety related reports and detailing their own experiences of near misses, unsafe acts and potentially dangerous situations. In order to successfully achieve this, a series of briefings, posters and awareness raising video campaigns have been performed. It is important that staff realise that their front-line knowledge and first hand experience and their involvement in operations is highly important and relevant for both management and other staff. In order to target staff groups, information sessions were performed by CIRAS staff for supervisory personnel, who then performed a series of cascade briefing until all staff had been introduced to the system. However, the under-reporting culture prevalent among railway staff (Clarke 1998) is well documented and traditionally staff do not fully trust managers. In order to develop trust in the system, CIRAS personnel also perform briefing sessions directly to front line staff. Further the system is confidential (although not anonymous), and the identity of the reporter is never revealed to managers. To date there have been no breaches of confidentiality. It has been demonstrated that feedback, anonymity and forgiveness are vital to ensure the success of any reporting system (Lucas, 1991).

Information on accidents, incidents and situations is collected via a report form or directly by telephone call to one of the three CIRAS centres. The initial report is followed up by a telephone or face to face interview. During the course of the interview, the interviewer determines the salient aspects of the report and discusses the nature of the concern with the individual; this addresses who, what, where, when, why and how questions. The interview is taped with permission, and the tape is transcribed to enable CIRAS staff to pass the knowledge gained onto the company. A questionnaire is also filled in at the time of interview, which collects basic demographic data plus other pertinent questions on environment, qualifications, consequences and recoveries. A full interview is absolutely essential as staff often write only minimal details on the reporting form. The reporting form itself simply asks for a description of the incident along with personal details. Initially when the system was first launched a detailed form with pre-defined tick boxes was used. However, railway staff did not find it easy to complete these forms and so a simple form was introduced.

Once the reporter has completed the follow-up interview, the form is returned and any personal details are removed from the system. However, the reporter's participation does not end when he or she has provided information. Individuals who make reports can receive feedback in two ways.

Firstly, reporters are provided with a unique reference number and can call CIRAS personnel following the elapse of a period of time and receive an update on the issue they have reported. However, this only provides information for the individual making the original report. It is perhaps a weakness of the system that individuals cannot be contacted by CIRAS once a satisfactory conclusion has been reached for a report: this is not possible as the original forms and reporter's details are destroyed or returned once the interview is completed (to maintain confidentiality).

Secondly, in order to share the information, experiences and problems encountered by staff at the front-line or sharp end, CIRAS produces regional journals detailing reports and company responses to those reports. This is distributed to employees' home addresses. In this way staff who would never normally know what has happened in another company (perhaps in similar conditions) can learn from the mistakes and experiences of others. In addition, each regional journal contains a national section that details reports which are relevant for the whole industry (e.g. rule interpretations). Anecdotally, staff enjoy reading the CIRAS journals and are interested by the problems encountered in other companies.

Knowledge Sharing

The information contained in CIRAS reports is dis-identified and conveyed to a designated contact within the relevant company. Each company has nominated a point of contact, who is in a senior enough position to decide on the importance of the issue and to pass it through the company channels in order to get answers. The information is relayed from CIRAS staff to the company representative in the form of face to face contact or via e-mail or letter. The regions maintain records of the information relayed and the resultant actions of the companies. In effect, this ensures that each company is provided with information on which to act or to build a profile of recurring incidents or problems. Individual companies are responsible for maintaining their own records of CIRAS events. The information gained is provided to each company and the company is then responsible for it's own action or lack of action. While CIRAS can make recommendations, it cannot force companies to take action.

Given the fragmentation of the UK railway industry since privatisation, it is often the case that a report concerns more than one company (e.g. a train operator and an infrastructure maintainer may be equally affected or responsible for incidents of near misses involving P-Way [track maintenance staff] and passenger trains). In cases like these the information is provided to all companies involved or affected. In this way CIRAS performs the function of sharing knowledge within the industry, which may otherwise be missed or lost in the system.

Raising Awareness

CIRAS aims to raise both company and individual awareness of hazards and risk factors via analysis of accumulated data and the use of the Journal. The Journal especially is a vehicle that is currently used by supervisors to bring dangerous or unsafe practices to the attention of staff during safety briefings and to discuss alternate ways of completing tasks or avoiding risks. In an industry where few actual accidents happen, the near misses reported via CIRAS can be invaluable in maintaining awareness among staff.

An outline of the system is provided in figure 10.1 below (activities within dotted lines are performed by Regions, other tasks are performed by Core).

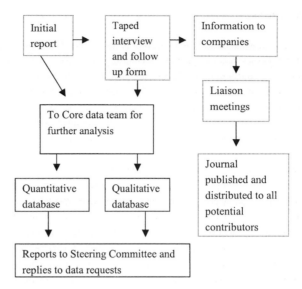

Figure 10.1: The CIRAS process

CIRAS APPROACHES TO KNOWLEDGE MANAGEMENT

There is much debate about the merits of personalisation and codification strategies for knowledge management. CIRAS uses both the personalisation and the codification strategies (Hansen, Nohria and Tierney 1999).

Personalisation

Within the CIRAS system the personalisation strategy takes three forms:

1. Eliciting implicit or tacit knowledge about safety from railway employees via reporting and interviewing
2. Providing individual companies' representatives with this information via discussions
3. Discussing the reports raised at a local forum with representatives from more than one company

The first has already been discussed. This section now turns to the latter two:

CIRAS staff extricate valuable information from the interview and pass this to individual company representatives. At this point the information is not codified into a set of root or underlying causes; rather the information as received from the individual (minus any identifying information) is turned into a situated contextual based narrative detailing the problem. These outlines of events may be discussed individually by company representatives and the CIRAS staff. This allows the individual representative to perform any audit or review of statistics or processes within the company prior to attending the Liaison Committee meeting.

CIRAS provides a forum where the companies get together with CIRAS representatives to discuss the reports from their various perspectives. This takes place at special Liaison Committee meetings. These groups meet every three months and allow the company representatives (usually senior safety managers) to discuss reports in depth, with CIRAS staff present. The groups operate at a local level. The function of the Liaison groups is twofold. Firstly, in these meetings the members discuss the problems that have been reported and provide some background information. They are provided with a summary of the reports prior to the meeting. The members discuss the underlying and root causes of the incidents, they discuss the likelihood of recurrence, their own experiences with this particular problem and determine whether the problem merits immediate action or whether the type of incident should be monitored. These groups discuss only individual incidents, while the National Steering Group is provided with amalgamated codified data. Secondly, at the Liaison committee meetings, the companies decide upon their response, and formalise it in a written statement which will be published in the CIRAS Journal: this states what action they are going to take and why. Where reports concern rule interpretations a reply is also sought from Railway Safety Ltd to prevent any ambiguity.

An example of this process is provided below:

Report: Train skidded on approach to XX (terminal station). Railhead contaminated due to passengers using toilet in platform.

At interview a number of other factors came to light that contribute to railhead contamination at this station.

Information passed to company: A driver encountered wheel slip on entering terminal station when travelling within the current speed limit. A number of factors appear to have contributed to the wheelslip. These are:

1.   *Toilets not tanked and empty directly onto rails*
2.   *Passengers and other public using toilets on trains when stopped at station (no signs prohibiting this or to suggest this causes a problem, also costs 20 pence to use public toilets)*
3.   *Leaky station roof adds to problem*
4.   *Oil from trains adds to problem*

Actions taken and reported in the Journal:
1.   *Weekly inspection instituted to determine extent of problem*
2.   *Biological trackmats laid to reduce the effluent*
3.   *Notices asking passengers not to use the toilets on trains when at a stand in station*
4.   *Station roof repairs already planned – date of commencement of work and timescale communicated to staff*

Codification

In order for a codification approach to support meaningful valid analysis, the data must be reliable. This is achieved by coding the data using a human factors framework and performing periodic inter-rater reliability trials. Reliability between coders is currently around 80%. In order to be beneficial to the individual companies the Core data team compile a six monthly report which provides in-depth information on the human factors problems, highlights common occurrences and trends and suggests ways to counteract or reduce the problems being reported. This report is discussed by an industry steering committee that meets every three months to oversee CIRAS and to decide on an industry-wide basis what actions should or could be taken as a result of trends emerging from the CIRAS database. The Steering committee comprises representatives of the industry who were elected to the position. The representatives are charged with cascading any decisions to the rest of the industry.

Furthermore, the data team also provides individually tailored company requests when received. This allows companies to compare the pattern of technical, organisational and human failures occurring within their company with the industry as a whole. As an example, a train operator may wish to know how frequently reports of communication failure occur and under which circumstances and may also want to know whether the pattern observed within his company is similar to the rest of the industry.

THE CLASSIFICATION SCHEMA

Human factors problems extend all the way from the operator to latent failures (see Reason, 1990). Thus, within the CIRAS system, human factors are broken down into three major categories – Proximal, Intermediate and Distal. Within each of these three categories, the CIRAS system breaks event reports down into a finer grained and more specific set of codes – these are shown in tables 10.1-10.3.

Proximal Factors

The CIRAS model includes skill based, rule based and knowledge based errors (Rasmussen, 1986 and Reason, 1990). These and additional 'sharp end' or proximal factors are shown below in table 10.1, with examples of the type of error made by the operator.

Intermediate Factors

Intermediate causes of events are defined as occurring between causes at the sharp end and organisational or systems causes (distal factors). Staff involved are somewhat removed in time and place from the occurrence of the event e.g. supervisors. Intermediate factors are shown in table 10.2 below, with examples.

Distal factors

Distal factors are defined as organisational or system related failures. Staff involved are likely to be remote in time and place from the event e.g. managers and designers. The distal factors are provided in table 10.3 below, with examples.

Table 10.1: Proximal factors

Proximal code	Definition	Example
Attention	Lack of concentration leading to slips (physical) and lapses (mental/ cognitive)	Failed to stop at station as a result of singing in the cab
Perception	Inability to see or hear specific features	Unable to see signal due to foliage
Knowledge	Lack of knowledge/ inadequate or incorrect knowledge for task	Trainee unaware that shunting procedure not authorised
Rule violation	Deliberate breach of rules or procedures	Not using electrical protection when necessary
Rule based error	Use of wrong rule/ procedure in a given situation	Using obsolete braking instructions
Skill based error	Inability to carry out a task	Driver has never stopped at correct point on a particular platform using viewing mirrors as misjudges distance
Fatigue	Tiredness/ fatigue influencing behaviour	Train delayed as driver got out of train to walk along platform as tired

Table 10.2: Intermediate factors

Intermediate code	Definition	Example
Communication between staff	Failure of communication between front line staff	Driver failed to pass on relevant information to signaller
Communication from staff to management	Failure of frontline staff to communicate with managers/ supervisors	Driver failed to report incident to supervisor
Communication from management to staff	Failure of supervisors to communicate with front line staff	Manager did not provide feedback on reported incident
Rule violation	Deliberate breach of rules/ procedures by supervisory staff	Briefing not given prior to work commencing
Maintenance	Inadequate or absent repairs/ maintenance	Monitors not maintained
Training	Insufficient for task, not provided	Training not provided on rarely performed task (isolating doors)

Table 10.3: Distal factors

Distal code	Definition	Example
Top down communication	Failure of senior managers to communicate with staff	Failure to fully explain implications of restructuring
Procedures	Ambiguous, difficult to follow or absent	Procedure surrounding inoperative AWS (automatic warning system) open to interpretation.
Design of equipment	Equipment design not fit for purpose	Lifting equipment for removal of train doors inadequately designed
Culture	Attitudes of work force, macho-style, general company ethos	Performance before safety

It is important that any coding system should be understandable to the managers who receive information from it. Therefore, CIRAS also has a classification system for technical/ equipment failures, specific to the railway industry, which includes in-cab/ on-train equipment and infrastructure hardware e.g. signalling systems. The system also classifies the personnel involved in the incident and the role of the public in contributing to the incident (e.g. trespass and vandalism, failing to stop at level crossings where there is no barrier control). Furthermore, the CIRAS system also provides information on the potential consequences of the near miss and importantly attempts to understand why the near miss did not develop into a more serious accident via a set of recovery codes. Recovery codes are only applicable to incidents (where a near miss has occurred) and not to issues (where the situation has the potential for a near miss or accident to develop). An overview of the classification scheme is shown in figure 10.2 below.

UTILISING THE KNOWLEDGE

The companies involved in the CIRAS system have used the knowledge gained via employee reports to make changes to their daily operation to increase safety and prevent a recurrence of similar incidents. These preventive measures should be monitored over time by the companies, the industry as a whole and the CIRAS system in order to gauge their effectiveness. As yet this is the weakest point of the system. In order to optimise national learning capacity it is important that the loop is closed at a national level. The first 18 months of operation has seen the system expand from less than 5,000 potential system users to 70,000. The next phase must be to establish responsibility for maintaining records of actions and to *ensure* that certain actions are carried out. It is currently unclear which body should have this responsibility: CIRAS, Railway Safety Ltd, HMRI (Her Majesty's Railway Inspectorate) or individual companies.

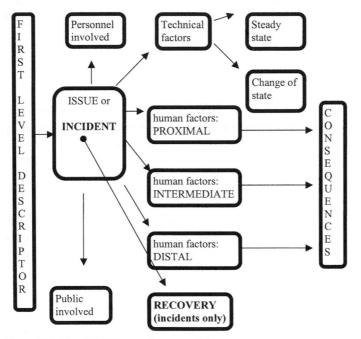

Figure 10.2: The CIRAS human factors model

The CIRAS data serve purposes at two levels. First, in the short term, a response is required to each report by the management of the company concerned, and a log maintained to keep track of issues which have been closed and those which remain to be resolved. Short term corrective actions are frequently taken where reports are shown to be well founded. Such actions include the replacement and repositioning of mirrors and monitors; changes to braking procedures on disc-braked sets; re-briefing and monitoring of electrical isolation procedures; apprehension by Transport Police of vehicle drivers failing to obey lights at two automatic level crossings; removal of foliage and visual obstructions; new lighting arrangements; changes to rostering;

fleet training day organised for infrequently carried out procedures; design of new lifting equipment for maintenance depot; erection of barriers at illegal crossing points; repair/ replacement of bleepers and communication equipment in a shunting yard.

However, in the longer term CIRAS offers scope for collecting generic human factors data that can serve as a resource for the industry as a whole. Currently ambiguous rules and procedures reported via CIRAS have led to national clarification and re-briefing (e.g the process for single line bi-directional working during a signal failure clarified).

DISCUSSION

Various aspects of the CIRAS system are discussed in the sections below, which are of relevance to any system that collects and disseminates knowledge within and/or across companies.

The scope of the system

It is important that any system, which aims to collect information that will be of benefit to the company, defines precisely the type of knowledge that is currently lacking and the information that it aims to capture. Information on near misses was lacking in the railway industry and CIRAS was established to fill this knowledge gap.

CIRAS does not replace the normal reporting channels within the various railway companies, but seeks to complement them. The CIRAS system concentrates on near misses, and small deviations rather than actual accidents. The railway industry already has systems in place to record accidents and incidents, but often fails to capture the near accident due to a perceived blame culture. Via CIRAS, information and knowledge, which may not come to light via the normal channels is collected, stored and most importantly shared within and across companies.

Administration

One of the strengths of CIRAS is that it is administered by an independent body. This in effect ensures that information is not used for commercial purposes i.e. the information reported to CIRAS has no direct effect on the awarding of franchises or contracts. As the administrators of the system are not connected to the railway industry directly, any information received and discussed with a company is not perceived as being related to anything other than safety and the promotion of safety. Commercial rivalry does not enter the frame as it would were a railway company running and administrating the system. With an independent body administering the system there is no perception of a hidden agenda by those who could report to the system.

Relationship with internal reporting channels and Her Majesty's Railway Inspectorate

Although CIRAS is administered independently it is not a replacement for normal company internal reporting channels. Rather CIRAS exists to collect data of a type that would not normally appear in official reporting channels (for example near misses, rule violations, errors and mistakes that did not result in an accident or incident). Since CIRAS was instituted within ScotRail Railways Ltd, there has been no decrease in the reporting to official internal reporting channels. It is important to clearly identify what information is useful to the reporting system and educate staff on the type of incidents that can be reported through CIRAS and those that should be reported through company schemes. Therefore a comprehensive training package is required for staff. For CIRAS implementation this comprised a video, literature and verbal briefing sessions. The method most appropriate to the company structure and resources should be used.

HMRI financially supported CIRAS during the initial stages of implementation and have the right to send inspectors to local Liaison Committees. They are furnished with a copy of the six monthly report and are kept up to date with results and progress made via the CIRAS system.

Management support

The success of any system is dependent upon the receptiveness of management to the information, and their willingness to act upon it. In order to gain the support of management joining the CIRAS scheme, a series of presentations to the boards of the various companies was performed. This allowed managers to fully understand the workings of the system and to have questions answered.

Within the railway industry there are examples of fully participative companies who use the system effectively to improve on their safety record and to identify problem areas. There are other companies who do not use the system fully, and often perceive it to be a time-wasting exercise because they have encountered similar problems before. However, this often means that the original problems were not dealt with effectively. While it is a strength of the system that problems are highlighted and discussed with the companies, the system is not in a position to effect action: companies are free to decide what to do with the information provided by CIRAS personnel. This is a major drawback of the system and should be addressed to close the organisational and industry-wide learning loop.

Data analysis

In a system that aims to highlight problems, it is vital that knowledge not simply be accumulated but also analysed. Without analysis recurrent problems and common causal factors will not be identified.

The data that are collected by the three regional centres are shipped to the Core data team monthly where the discourse is coded according to the CIRAS human factors model. This model has been specially tailored to meet the requirements of the railway industry, as well as to include operator and organisational codes. The Core software includes a text searching facility that allows similar instances of events to be selected. In fact, each of the technical, proximal, intermediate and distal codes assigned to a report is associated with text from the interview in a special database for content analysis. The ability to search and retrieve text in a large reporting scheme is crucial, although problems of recall and precision are faced by most relational databases and search engines (Johnson, 2000). Text relating to particular topics is stored at an appropriate node by data analysts. Text can be retrieved individually or along with all other text stored at that node. Hence transcripts may be searched for information on a particular topic (e.g. Signal Passed at Danger) or the node where information is stored may be searched (e.g. knowledge).

The system offers the service of a tailor made analysis for the individual companies. A number of companies have taken advantage of this, although these tend to be the more open and receptive companies. In this way, CIRAS information can be added to the knowledge and intelligence already collected by the companies.

The six monthly report on trends in the data and a detailed analysis of the data represents a real gain for the industry. This is an additional resource for gaining knowledge about certain aspects of the industry – for example CIRAS data has already fed into an industry wide analysis of obscured signals and communication failures.

Staff support and motivation to report

The success of any technical system for gaining and sharing knowledge also rests on the trust and participation of staff at the sharp end who are in the best position to know how "it really works at ground level". CIRAS has made it possible for sharp end staff to make reports without the fear of discipline or retribution. Support of the front line staff has been gained via briefings on the working of the system, question and answer sessions performed by CIRAS personnel, management action on reports and an interview process that is designed to obtain any details which would inadvertently identify a reporting individual. The interview process goes beyond merely examining often poorly written or incomplete reports, and allows a full account of the incident to be relayed. Staff become more motivated to report to CIRAS when they see real changes and improvements being made based on the reports of themselves or colleagues. Through this process the knowledge of the way near misses arise has increased. The CIRAS process has also helped to close the gap between how procedures work in practice and how they ought to operate in theory.

Feedback for staff

Many reporting schemes are seen as 'black holes'. The main reason for the railway under-reporting culture identified by Clarke (1998) was that "managers take no notice" and "nothing would get done". In order to maintain the support of staff a scheme that collects knowledge and information from front line staff, should also provide those staff with feedback.

At a local level, the quarterly CIRAS journals demonstrate to staff that their reports are important and are taken seriously by the company. The journal also enables staff in different geographical locations and in different companies to understand problems which may be similar to their own. The journal has also made drivers and signallers (two groups which traditionally dislike each other) more aware of the difficulties and problems encountered by the other. Issues of national importance are identified by the Core data team, and are also included in the journals.

Knowledge sharing within and between companies

In order to maximise learning and knowledge sharing, Liaison Groups meet every three months. This provides an important forum where reports are openly discussed by multiple companies and remedial actions are decided upon. This forum ensures that all parties have access to the different incidents that have been reported and can learn from the adverse experiences of other companies. Such a forum is essential to the success of the system and maximises the knowledge sharing and organisational learning of the system.

REFERENCES

Clarke, S. (1998). Safety culture on the UK railway network. *Work and Stress* **12** (1),6-16.

Davies, J. B., L. Wright, E. Courtney and H. Reid (2000). Confidential incident reporting on the UK Railways: The 'CIRAS' system. *Cognition, Technology and Work*, **2**,.(3).

Hansen, M. T., N. Nohria and T. Tierney (1999). What's your strategy for managing knowledge. *Harvard Business Review*, **28** (2), 65-71.

Heybroek, R. (1995). *Improving safety training: Human factors discrepancies report.* Vosper Thornycroft (UK) limited, MSC Division.

Lucas, D. A. (1991). Organisational aspects of near miss reporting. In: *Near miss reporting as a safety tool* (T. W. van der Schaaf, D. A. Lucas and A. R. Hale, eds). Butterworth-Heinemann Ltd., Oxford.

Johnson, C. (2000). Supporting the analysis of human error in national and international incident reporting schemes. In: *Proceedings of EAM 2000, 21ˢᵗ European Annual Conference Human Decision Making and Manual Control (* P. C. Cacciabue. Ed.).

Rasmussen, J. (1986). *Information processing and human machine interaction.* North-Holland Amsterdam.

Reason, J. (1990). *Human error.* Cambridge University Press, Cambridge, UK.

Van Vuuren, W. (1998). *Organisational failure: an exploratory study in the steel industry and the medical domain.* (Ph.D. Thesis, Eindhoven University of Technology).

Wright, L. B. (2000). *The analysis of UK railway accidents and incidents: a comparison of their causal patterns.* (Ph.D. Thesis, University of Strathclyde).

PART 3

LEARNING FROM SUCCESSES

11

COMMUNITIES OF PRACTICE FOR KNOWLEDGE SHARING

J. H. Erik Andriessen, Mirjam Huis in 't Veld and Maura Soekijad[1]

INTRODUCTION

Management of knowledge is one of the most discussed subjects in recent organisation science. Two developments have contributed to this phenomenon: the change from industrial to service (=knowledge processing) organisations and the growth of information and communication technology. The first development has necessitated an intensive study of the processes around the core assets of most modern organisations, i.e. of information and knowledge. The second development has lured many organisations into believing that sophisticated ICT tools are the answer to questions concerning the handling of information and knowledge. The whole 'knowledge value chain' (e.g. Weggeman, 1997), i.e. acquiring, developing, storing, distributing, applying and evaluating knowledge, was considered to benefit enormously from investing millions of dollars in 'knowledge technologies'. Procedures to elicit knowledge from employees, converting it into a systematised form and storing it in company wide repositories are the core activities of what is labelled the ***codification strategy*** (Hansen *et al.*, 1999). Information storage, retrieval and exchange can indeed benefit strongly from digital information systems. Knowledge, however, is different from simple information. It is information that is experienced and interpreted by a person, it is related to an actual situation and it makes sense to that person. Knowledge is often very implicit, not consciously

---

[1] Delft University of Technology, The Netherlands

articulated, i.e. it is tacit. Strictly speaking when knowledge is explicated into a data-system it is no longer contextualised and personally sense making, so it becomes information again. But the difference between information and knowledge is not sharp, it is rather a scale with grey zones: some information, such as lists of names, is hardly more than data, and some information, such as stories concerning the way a certain person has solved a particular problem is almost knowledge.

The codification approach with regards to *knowledge* has seen many failures. The reasons are diverse, but the most important ones are that much knowledge – in the sense of personal experiences – is very difficult to explicate, and secondly because of the psychological resistance against providing and against using knowledge that is separated from its owner, i.e. that is made *impersonal* (see Huysman in this book). Exchanging knowledge with others may provide status and is trustworthy for the receiver. Putting knowledge in a system is cumbersome, removes it from its context and rarely provides personal rewards.

A second way of dealing with sharing, applying and developing knowledge in organisations is what Hansen *et al.* (1999) call the *'personalisation strategy'*. In this strategy the focus is on people meeting each other, on interpersonal knowledge sharing, on master-apprenticeship relations, knowledge intermediaries ('knowledge brokers') and on what is called 'communities of practice' (see below). The role of ICT applications is limited to communication systems, supporting the interaction between people, and to e.g. 'yellow pages', i.e. information on which experts to find where. The authors argue that the first strategy is called for in organisations (or departments) where the information processes are rather routinised, e.g. consultancies with more or less standard projects. The personalisation strategy is considered to be appropriate for organisations (or departments) with non-standardised, i.e. creative, novel production processes. This argument however is debatable. Some organisations have very adequate combinations between knowledge systems and personalisation initiatives such as communities or knowledge intermediaries. De Bruijn and De Neree (2000) make a convincing plea for an approach in which these competing values are combined.

The arguments concerning the two approaches for knowledge management can be fruitfully re-interpreted in terms of organisational learning theories (see also e.g. chapter 2, 3, 5 and 7 of this volume, see also Huysman and De Wit, 2002). These theories imply that learning takes place not simply by individuals, but also in group and organisational context. Individuals learn from outside sources such as clients and conferences (short incoming arrows at the top left) and from inside sources, i.e. from organisational knowledge systems and from personal activities,

In this process of internalisation (in terms of Nonaka and Takeuchi, 1995), *integration of various types of knowledge*, but also *problem solving* and *experimenting* play a role

Individuals share their knowledge with others, which may result in shared (group) knowledge. This can be explicit knowledge exchanged through e.g. presentations, or implicit, tacit knowledge exchanged through working together ('socialisation'). When shared knowledge is accepted by the organisation it becomes organisational knowledge, which is than available to be distributed again to individuals or groups. Both individuals and groups may store knowledge in (explicit and digital) organisational repositories, or in the implicit norms, values and knowledge of the organisation, i.e. its culture. These storages then form the organisational memory. Finally, knowledge is exported from the organisation by employees participating in external groups and meetings or by leaving the organisation.

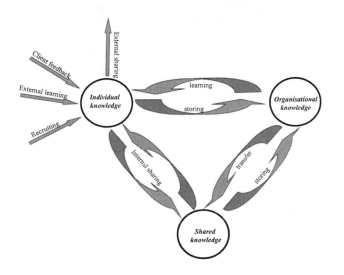

Figure 11.1: Knowledge processes in organisations (adapted from Huysman and De Wit, 2002 see above)

The interaction and knowledge sharing between individuals can take place in dyads, but also in groups. Specific groups, aimed exclusively at knowledge sharing and developing, are often called *Communities of Practice (CoPs)*. They form a framework, by which both explicit and tacit knowledge can be shared, both for solving individual or common problems, and for developing new knowledge. Of course, communities are not the only vehicles for this transfer of knowledge. Ad hoc meetings and encounters, master – apprentice relations or knowledge intermediaries may also realise this objective. But the last few years many modern organisations, and networks of organisations, are experimenting more or less formally with CoPs. They feel themselves heavily dependent on knowledge sharing, since the short life cycle of project teams, the geographic distribution of expertise and rapid employee turnover may easily result in reinventing the wheel again and again. Connecting employees together in more or less permanent communities may solve this problem to some degree, although it has to be complemented by integrating the knowledge, generated in the communities, into the organisation as a whole. The ideas of Argyris and Schön (1996, see also chapter 2 in this volume) suggest that CoPs should therefore have a double function, i.e. of facilitating the interaction of individual members, and of bridging the gap between experience sharing individuals and the organisation. According to the theory of Argyris and Schön, organisational learning from individual experiences requires an agent assigned to learn for the organization. This agent can then function in 'cooperation' with a knowledge system, firstly to 'translate' the individual experiences into innovative solutions for the organization ('double loop learning') and secondly to care for the proper storage of the solution into knowledge systems, i.e. training material, manuals or digital repositories. In this way the above-mentioned antagonism between the codification and personalisation approach may also be solved.

In this contribution we focus exclusively on Communities of Practice. We will deal with two basic questions:
1.   What kind of communities of practice can be found?
2.   What are the conditions for the successful functioning of communities?

These questions will receive preliminary answers on the basis of literature, some case studies of CoPs, and through the application of a group dynamics model that was developed to study various types of groups.

THE CONCEPT OF COMMUNITIES OF PRACTICE

Researchers from Palo Alto Research Centre and the Institute for Research on Learning started, in a sense, the 'knowledge revolution' in the 1980s. Lave and Wenger (1991), together with others, developed new concepts to understand how learning occurs and knowledge becomes shared within communities. They showed that often the best kind of learning is 'situated learning', where social interaction is critical. Formal training settings may be necessary to transfer basic information, but for real learning, practice oriented interaction with a trusted group of colleagues is essential.

Lave and Wenger (1991) focused on the study of certain craft communities such as butchers. With the concept of 'Legitimate Peripheral Participation' they pointed at the relation between newcomers and old timers in those communities. Newcomers are legitimate members of a community, but they participate at the periphery. In the centre are the more experienced members of the practice (experts). Through observing, imitating and co-operating the newcomers slowly learn all the elements of the practice and become experienced members themselves. In this way the community reproduces itself as a living entity.

Brown and Duguid (1991) studied the relation between formal and informal organisation. Many organisations are willing to assume that complex tasks can be successfully mapped onto a set of simple Tayloristic, formalised steps that can be followed without need of significant understanding or insight, and thus without need of significant investment in training or skilled technicians. In actual practice however, work processes are often deviating from the formal prescriptions, because these prescriptions may be outdated, or too rigid and not applicable to exceptions. The burden of making up the difference between what is formally described and what is actually needed then rests with the employees, who share the non-formalised knowledge and in that way actually protect the organisation from its own short-sightedness. The authors conclude that an organisation may probably best be structured around knowledge and practice in communities in order to improve its innovative potential.

Central in these ideas is the concept of 'practices', around which communities build, acquire, and create their knowledge. This is not theoretical knowledge, but knowledge related to a common practice such as a professional discipline, a skill or a topic. Sharing knowledge in a community will build or enrich a (set of) common practice(s). A community builds capability in its practice by developing a shared repertoire and resources such as tools, documents, routines, vocabulary, stories, symbols, artefacts, and heroes that embody the accumulated knowledge of the community. This shared repertoire serves as a foundation for future learning (Allee, 2000).

The idea of communities as breeding grounds for sharing experiences and particularly for developing innovative concepts has found an open ear and eye in modern organisations., These organisations are looking for ways and means of strengthening its most important asset, the knowledge embedded in their employees, who are often distributed world-wide. Instead of referring to processes of knowledge transfer to newcomers in small groups of co-located craftsmen, or to the struggle of informal shop floor workgroups, the concept of 'communities of practice' became related to groups of professionals, generally from different organisational units and geographically distributed, that share a common interest in a certain knowledge domain. McDermott (2001) defines them as "... *loosely knit groups driven by the value they provide to members, defined by the opportunities to learn and share what they discover and bounded by a sense of collective identity"*.

Another reason for the growth of CoPs is the fact that many organisations are project and market oriented. The traditional functional departments, where specialists of the same discipline were working side by side, have made room for product-oriented groups where, however, good mechanisms for inter-project knowledge sharing between experts often are not available.

CoPs can be distinguished from teams in the sense that teams are formally institutionalised with the objective of achieving a certain product within a certain period of time. Communities of practice are loosely coupled groups that provide their members with opportunities to learn and to share experiences. Teams progress by moving through a work-plan, while communities develop by discovering new areas to share current knowledge and develop new knowledge around the practice. That is why managing these two groups is a very different job. Managing a team consists of co-ordinating interdependent tasks. Managing a community implies making connections between members and keeping the topics of the community fresh and valuable (McDermott, 2000).

In actual practice the differences between teams and communities are often not that sharp. The practice and the literature on CoPs show a wide variety of forms. The discussion is complicated by the fact that different names can be found for concepts or constellations that show similarities to CoPs, such as virtual communities, knowledge communities, and occupational communities. Virtual communities can be found not only in organisations but also on the Internet, focussing on a certain interest (e.g. hobby), on political issues (discussion platforms) or social issues (e.g. the arthritis community) (see van den Boomen, 2000). Many communities in organisations do prefer a certain degree of face-to-face meeting, but, as will be illustrated below, the *degree of virtuality* can differ widely. ICT facilitation is central in the study on virtual communities (e.g. Hildreth, Kimble and Wright, 2000).

FIVE EXAMPLES

In this section we will describe shortly the way several large companies are dealing with CoPs. The information is based on case studies by ourselves (Unilever, Atos Origin, Delft Cluster) and on contacts with key informants of the other organisations (Andriessen *et al.*, 2001). More information on these cases will be presented as illustrations in the following sections.

Unilever

The Knowledge Mapping and Structuring Unit at corporate level has started initiatives such as 'Knowledge Workshops' and 'Communities of Practice' (CoPs) to enhance the efficiency of production and to improve innovative processes. In these initiatives, both knowledge exchange and knowledge creation processes play an important role. The first knowledge workshop was organised when the company faced problems in the processing of tomatoes. Several experts concerning the production of tomato-products were brought together. As an unforeseen phenomenon the workshop gave birth to a community of experts, since the participants, having met and learned from each other quite intensively, kept contact and exchanged new ideas after they went back to their own companies (Huysman and De Wit, 2002). At this moment the practices of the communities at Unilever concentrate mostly around certain production processes (e.g. of margarine and tomato products), but some CoPs are also to be found concerning principles of supply chain management and quality norms.

Setting up communities at Unilever proceeds quite formally. A high level management 'champion' is committed and together with him ten to twenty organisationally and geographically distributed employees are selected carefully and then asked to join the community. When they accept to join each receives a special status in his or her own department. From that moment on they are considered to be representatives of their department in the community and vice versa. The experts are brought together for a workshop of about a week, to exchange information, they organise the group and care for teambuilding. A facilitator co-ordinates the group activities. The corporate Knowledge Structuring and Mapping Unit developed a handbook for facilitators. The community is expected to communicate continually, but in many cases the most intensive distributed communication occurs shortly before and after the face to face meetings that take place once or twice a year.

Shell

Shell is an oil company of Dutch - British origin. The organisation is divided among the three basic businesses of oil, chemicals, and exploration and production (E&P). The 'division' of E&P has ca. 30.000 employees, of which about 70% is member of some kind of network.

In 1998 Shell contained many small communities of 20 to 300 members. The groups were mostly informal in origin, with hardly any structure or facilitation. In 1999, the small groups were combined into global networks called communities of practice. These groups are quite large and have a formal position in the organisational structure. In E&P communities can be found on the issues of sub-surface processes, of surface processes and of wells. Each CoP has 1500 to 2000 members. Smaller communities are dealing with issues of e.g. competitive intelligence or of Human Relations. The communities have so-called 'hub-co-ordinators' for facilitation. They meet each other once in three to four months. They are responsible for the co-ordination of all activities within the various communities.

The role of most communities is limited to daily problem solving. They serve mainly as a source of information for those members who have a problem in their work and seek the expertise of colleagues to solve this problem. Embryonic subgroups may form for a short time, discussing a specific issue. Members do not meet face-to-face, but send their questions and reactions via a simple email discussion list facility. This means that the learning that takes place is single loop learning, rather than double loop, innovative learning (Argyris, 1992, see also chapter 2 of this book). However, the department responsible for working standards regularly analyses the email messages to find elements that may be turned into standards. In this way shared knowledge is turned into organisational knowledge.

BP Amoco

BP Amoco is the third-largest oil and gas company in the world. The organisation was formed at the end of 1998 from a merger between British Petroleum (BP) and Amoco. It consists of 126 business units in over 100 countries and employs over 80.000 people. Knowledge sharing was originally organised through formal networks, in which people meet regularly face-to-face. Later, a large number of informal networks (CoPs) began to grow, sometimes on the basis of already existing loose affiliations of practitioners. The CoP-concept then helped to legitimise them and gave them direction and impetus. Examples of communities are those on '3D reservoir modelling', 'production efficiency' and 'water produced during drilling'. Since the merger with Amoco, the organisation has developed the so-called 'dual (formal - informal) network model' (Collison, 1999). The differences are shown in the following table.

Table 11.1: Communities at BP Amoco

	Formal Networks (Communities of Commitment)	Informal networks (Communities of Practice)
Membership	Defined	Voluntary
Customer	Business customer or peer-group	Members are their own customers
Collective performance contract	Yes	No
Management process	Formal	Informal or none
Level of company resourcing	High	Often low or none
Examples: communities of	Maintenance managers; operations managers; peer-groups	Seismic survey; problems of grease; produced water; 3D-modelling network
Number of networks present in the organisation	50 - 60 CoCs	250 - 300 CoPs

There is no defined hierarchy in the communities of BP Amoco, but in practice certain experienced members answer most of the questions at discussion groups and are active in steering and summarising discussions. Approximately half of the community-members participate in such discussions, while the others do not.

Because most communities do not meet face-to-face they rely on ICT. The tools used at BP Amoco include systems for email, public folders, discussion groups and shared documents. The networks can use various means for facilitation, such as a network facilitator (who co-ordinates the activity of the network), several means of communication (both face-to-face and virtual), a common storage facility for community knowledge (LINK-tool), and the Connect tool. Connect is an Intranet tool that serves as Yellow Pages, i.e. it contains information regarding the company members and their expertise. It is linked to desktop videoconferencing, multimedia email, and contains links to business units, teams, networks and external Internet sites.

Two things proved very important in this BP Amoco case. First, the tools were developed in direct relation to the employees and their needs. Second, although the use is voluntary, people are offered a suitable training to use the tools properly.

AtosOrigin

AtosOrigin provides ICT services including consultancy, implementation and system integration. These services are provided world wide, with a total of 28,000 staff. Six thousand staff are based in the Netherlands in various geographical locations. Within AtosOrigin in the Netherlands, there are three types of communities, i.e. (local) 'Expertise Groups', (national) Networks of Performance and (national) Performer Groups.

Expertise Groups are initiatives within (Dutch) regional sections of the AtosOrigin company. They consist of consultants working in that section, who exchange experiences concerning work related topics. Examples of the topics are Oracle Databases, Microsoft software, Java, but also Project Management. They come together face to face about once a month. Each consultant in that section of the company has to be member of at least one EG and – in some regions - can be a secondary member of one or two others. The use of ICT is very limited.

Networks of Professionals are bottom up growing networks of company employees, distributed nation-wide. The goal of NoPs is not only to exchange but also to create new knowledge. The topics of the NoPs are to some extent parallel to those of the expertise groups. NoPs have to have at least five members in order to be granted official status and be given company backing and support.. There are no rules relating to membership or functioning of the community. Individuals may be member of more than one such group. The members interact through various ICT tools, but may also organise face-to-face meetings.

Performer groups are particularly focused on developing and storing best practices and guidelines for project management in certain domains. The storage is done in a dedicated software tool called Performer. . The groups cannot be established without management approval, and employees wishing to join a Performer group must formally request membership via the group moderator. Anyone who is a member of a performer group can add information to the shared database, although the moderator has the final decision on whether items added to the database should be altered or removed.

Delft Cluster

Delft Cluster (DC) is a consortium of five organisations in Delft, the Netherlands, that focus on consultancy and research concerning sustainable river-delta development. The consortium consists of research institutes and companies. This network of organisations was formally established in 1999 and is sponsored by the Dutch government. DC has defined seven themes of expertise and each of these themes has defined several projects in which interested sector organisations can participate. The projects contribute to the overall DC goal to strengthen its knowledge and position in the field of sustainable river-delta development. The five organisations have a common goal of developing and sharing knowledge about river-delta development, but at the same time they can be competitors when trying to acquire commercial or scientific projects. Members of the organisations meet each other in knowledge sharing communities but also in various kinds of other interactions such as research programmes, projects, consortia or advisory boards. The DC cases give therefore many examples of 'overlapping memberships'.

The communities have varying practices concerning interactions: formal meetings, informal gatherings connected to other meetings such as of committees or at conferences, or even only contacts between some members when an urgent issue arises. The role of ICT is very limited. Most people involved have there own email facilities, but the communities as such are hardly supported by proprietary systems. DC has tried to introduce a groupware tool BSCW for all communities to use, in order to overcome incompatibility of systems used. The introduction of this tool, however, proceeds only very slowly.

TYPES OF COMMUNITIES

The above-mentioned case descriptions indicate that the concept of Communities of Practice is applied to various types of groups. They have a few characteristics in common, which can be used to formulate a definition of CoPs: *Communities of Practice are intra- or organisational groups that are formed with the aim of sharing knowledge around a certain 'practice'. They are long term oriented and have loosely coupled membership, but share group identity.* This definition excludes Internet communities, project teams or professional associations. Nevertheless CoPs may differ along many dimensions, amongst which the following are quite central:

1. *purpose*: individual learning and daily problem solving (single loop learning), organisational innovation (double loop learning), or networking for finding knowledge.
2. *formalisation:* : top down formally initiated, with centrally selected members and an appointed coordinator, or informal, spontaneous, bottom up, with emerging 'leaders'
3. *size*: from small to very large
4. *boundary*: open or closed for people inside or outside the organisation
5. *composition*: only experts or experts plus newcomers

6. *virtuality*: high (not meeting face to face, communicating via ICT) or low (meeting mainly face to face)

Certain combinations of these dimensions are rather frequent, resulting in the following five types of CoPs:

a. The *'Daily Practice community'* (e.g. AtosOrigin's expertise groups), consisting of experienced workers and newcomers, often working in physical proximity, coming together to discuss daily experiences;. These groups resemble the original craft based 'communities of practice' described by e.g. Lave and Wenger (1991).

b. The *'Formal Expert Community'* (see Unilever's CoPs, BP's 'formal networks'): a limited number of geographically and organisationally dispersed experts, rather formally instituted as group, with the purpose of exchanging and/or developing knowledge in a certain domain; interaction may occur via ICT, but face to face meetings still play an important role. The difference between this type of CoP and project teams may be rather vague.

c. The *'Informal Network Community'* (e.g. BP's CoP's, AtosOrigin's NOPs): an informal, freely accessible, group of people, formed to discuss a common issue of interest. They are generally geographically and organisationally dispersed and communication is often exclusively through the ICT media.

d. The *'Problem Solving Community'* (e.g. Shell): consisting of all (geographically and organisationally dispersed) employees of the same function, such as all oil drillers. Through the ICT network they exchange questions and answers concerning the solution of certain practical problems. Although the size may be very great, they still display some form of group identity, based on commonality in function and organisation, and fall therefore under the definition of CoPs.

e. The *'Latent Network Community' (e.g. Delft Cluster)*: A group of persons, often working in different organisations, who know each other well, but interact mainly in other settings. The group as such rarely comes together. In some periods these networks may be quite similar to type b (formal experts) or c (informal network)

Except for type a, these types of CoPs may be found both within organisations and between organisations.

The most central dimensions along which CoPs appear to differ is that of purpose or function. Functions are similar to success criteria. The question is when CoPs are considered to be successful, and the answer to this question appears to differ for the various types of communities and also for different stakeholders involved. Some CopS are considered to be successful when the individual members can learn from each other, or when their social function is strong. Other CoPs are expected to produce certain products for the company. This indicates that success of communities may be defined in terms of the three types of functions all groups have, although in varying combinations (e.g. McGrath, 1984):

- the *production* (organisational) function: a group is successful to the extent that it contributes (through its output) to the effectiveness and innovation of the organisational context. For communities this function is reflected in the collection of best practices, in the development of innovative ideas or simply in keeping the knowledge of job hoppers in the company.
- the *group vitality* function: a group is successful to the extent that it develops trust, and identity. For communities this function is expressed in the group cohesion and motivation this provides to their members
- the *member support* function: a group is successful to the extent that membership is rewarding for the individual group members. For communities this is reflected in the possibilities to stay up to date with the latest developments concerning the CoP topic, but also concerning the organisation, to find experts and to find information much faster than through other channels.

Individual members of communities are only motivated to exchange knowledge if they perceive strong personal incentives. The question is to what extent the other two functions should also be present. The production function appears often to be quite debated. In some organisations the Return-on-Investment question is explicitly asked: CoPs have to contribute to the organisation's effectiveness, preferably through providing clear financial benefits, or at least through the development of best practice manuals or other specific products. In other organisations the communities are expected to contribute to the organisation in an indirect way, i.e. through the learning of the individual members.

Indeed, the primary function of communities is not to perform a specific task within deadlines, but to exchange and sharing knowledge, with the aim of individual and collective learning. However, if organisational learning has to take place, the knowledge shared in the community should be made available to the wider organisation (see introduction to this chapter). Communities may assume the role of 'agent for double loop learning' (see introduction) and translate the shared experiences into overviews of best practices, manuals or innovative ideas. The central issue is, to which extent this role should be enforced.

At Unilever, communities that cannot demonstrate clearly its benefit for the organisation are threatened to be dissolved. Or members can no longer get the possibility to join, because their direct chief would not give them time or money anymore. At AtosOrigin however it is believed that the main benefit for the organisation would arise through the improvement of the level of knowledge of the individual members. Nevertheless, also in this company communities were encouraged to produce e.g. overviews of best practices. And various regional expertise groups, who all dealt with a similar topic, had a representative in a national community to exchange the knowledge that was developed in the regions. Through this channel the regional knowledge was made available to the company as a whole

Summarising, the main issue is not whether CoPs should contribute to organisational learning. Most of them do, be it in a more or less formalised way. The issue is that companies should not approach this from a perspective in which communities are considered as a kind of hobby of their members, but from a perspective of necessary elements in a strategy of organisational learning.

CONDITIONS FOR THE SUCCESS OF COPS

Success - in terms of individual learning, group vitality and/or producing organisationally relevant products - is a function of many conditions. To distinguish the various types of conditions we developed our version of a well known input-process-output model, on the basis of many studies of group dynamics, effective teams and groupware appropriation (e.g. Hackman, 1987; McGrath, 1984; Anderson and West, 1998; Campion *et al.,*1996; Guzzo and Dickson, 1996; Poole and DeSanctis, 1990). The model is called the *Dynamic Group Interaction Model* since it explicitly takes into account the learning and adaptation stages of group development (see Andriessen, 2002). A shorthand version is visualised in figure 11.2.

According to this model the success of groups depends firstly on the way group members interact (five processes, plus feedback, are distinguished), and secondly on the characteristics of the setting, i.e. characteristics of the individuals, the group, the tools and the environment. These characteristics however are not static but can change, particularly in the early stages of a group. Theories of learning, of structuration and of appropriation deal with these feedback processes. The interaction processes in a group change the context-characteristics of the group, thereby leading the group through certain 'life cycle stages'. In this dynamic perspective context characteristics such as the group task or trust and cohesion are both conditions for and output of group processes, depending on the moment of observation.

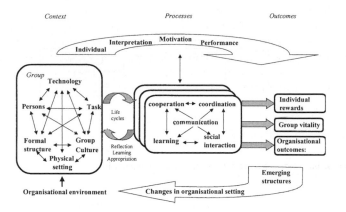

Figure 11.2: Dynamic group interaction model

More precisely, the success of a group depends on the degree to which the processes mutually match to each other, the degree to which the input characteristics match each other and the extent to which learning processes function adequately. For communities this means e.g. that the formal structure should fit the goals of the community. Since the goals could be related to various outcomes, they may be difficult to reconcile: continuity of the community is probably served by strong formalisation and institutionalisation, while spontaneous knowledge sharing may be better served by weak formalisation. In the same way the processes may be inherently incompatible: strong co-ordination may be difficult to match with knowledge creation and learning processes. Certain generic conditions seem to hold for many CoPs (see figure 11.3).

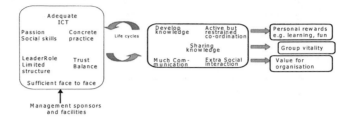

Figure 11.3: Conditions for successful Communities of Practice

However, the purposes and therefore the role of conditions are different for the five CoP types mentioned above. Problem solving CoPs appear to have quite limited purposes and can therefore thrive without much conditions, except a good email system. The success of formal expert communities however requires much money and support. Informal network communities need a very active coordinator or core group and adequate ICT support, while Daily Practice communities require passionate members and coordinator.

The model implies that the study of CoPs should focus on the one hand on the actual processes that take place and particularly on the learning processes and how the other processes are tuned to this central process. On the other hand the role of context factors are very crucial, both the formal and the informal aspects. In the following section a few paradoxes in this respect will be discussed, originating in the case studies we have performed .

DILEMMA'S

The matching requirement appears to result in several dilemmas that communities face. Insight in the dilemmas is very fruitful in the discussion concerning conditions for success.

Formalisation or spontaneous growth. Successful project teams – having clear tasks and deadlines - require strong formalisation in certain aspects such as role division, planning, and membership and also in the day to day coordination processes. Organisations often apply this team logic to communities. Communities are then viewed as knowledge innovation 'tools'.

Tools have to be controlled and that is what many companies do. This control can serve two functions. Firstly the provision of clear results, secondly the continuity of the community. This implies that those organisations are keen on controlling the selection of the people who will join the community and the roles they are going to fulfil, controlling the way the members of these communities interact and, most important, controlling its deliverables.

At Unilever and AtosOrigin a relatively high degree of formalisation can be found, especially in the phase of preparation and set-up of communities. Employees are invited to become a member of a community when the knowledge object is part of their daily work practice and when they are considered to be experts in the domain.

Organisations, or at least executives, who have to accomplish targets, tend to focus here on short term, tangible deliverables. And they also expect that continuity is more probable, when the community is institutionalised and supported centrally and formally, with clearly established roles and procedures.

However certain literature on communities appears not to favour formalisation. To be innovative and creative, communities need room to explore the field, members like spontaneous bumping into some enthusiastic peers and they need informal brainstorming with a long-term dimension. These different desires may create strong tensions. Although communities are often self-organising and thus resistant to supervision and interference, they sometimes need specific managerial efforts to develop and integrate them into the organisation. (Wenger and Snyder, 2000)

External versus internal connection: The organisational context forms an important condition for the successful functioning of groups. According to some authors (e.g. Sundstrom *et al.*, 1990) boundary-spanning processes and the interaction with the organisational context are even the most important success-conditions for teams. Many teams appear to fail in that way since they are too introverted. Communities however seem to face the opposite danger. The lack of a common task may lead to very low cohesion and binding, and a strong pull by the members' own department.

In almost all communities the individual members have strong relations with the environment, such as their own boss and company-unit. Maintaining a good relationship with that context appears to be crucial both for the individual members as for the community as a whole. The demands of their boss and work-environment prevent community members sometimes from spending much time on the community. To cope with this problem rules may be very clear as to the importance of participating in CoPs and concerning the fact that members are (annually) evaluated as to there level of participation in a CoP (see e.g. AtosOrigin).

Competition or cooperation: Related is the tension between 'giving to' and 'taking from' the community. CoPs flourish only when members have the feeling that they get something out of the community, i.e. when the community is personally rewarding. However, if a member is focused on getting out as much as possible for him / her self and does not have the intention to share his/her own knowledge the co-operation is likely to fail. A specific form of this dilemma is that of competition vs. co-operation. When individuals in a community originate from various organisations that are potential competitors, they can encounter a coopetitive dilemma (Brandenburger and Nalebuff, 1996). On the one hand they can feel competitors, which might result in the strategy of withholding knowledge since 'knowledge is power'. This might cause difficulties in the sharing and creation of knowledge. On the other hand, these participants can perceive each other as value adding actors that complete instead of compete each other. Than they will more easily apply a 'knowledge is strength' strategy in co-operating in the community. Trust has an important function in this latter strategy.

Loosely or tightly coupled: The central characteristic of Communities of Practice is the sharing of knowledge. While the issue of shared knowledge receives less attention in the team literature, it appears to be of major importance, to be the 'raison d'être', for communities. Sharing knowledge is both the objective of but also a condition for successful functioning of communities. Sharing knowledge appears to be a subtle process that requires a climate in which people are motivated to exchange information and (tacit) knowledge. A major condition for these processes to taken place is trust (Kramer, 1999; Klein Woolthuis, 1999) and the question is to what extent communities should explicitly develop group identity (teambuilding).

Sometimes trust has to be build through explicit team building activities. For the Unilever formal expert communities this appeared to be an important process But for the problem solving purposes of the large Shell communities the feeling of common identity existing among the 2000 Shell oil-drillers is sufficient for a successful exchange of questions and answers.

ICT: low or high profile, customisation versus uniformity: Communication and co-operation, but also document exchange and information retrieval in geographically distributed groups may be supported by ICT. Despite many studies however, little is yet known about the role

Collaboration Technology can fulfil in support of effective teamwork, let alone for communities. In practice one can find extremes as to ICT support.

BP Amoco has a very sophisticated set of tools (see above), while in the Shell communities the IT toolset in use is quite limited,. The most frequently used tool in Shell is a discussion group facility on which everyone can post a request or question for help, while others react to it. Currently, there appears to be no need for a storage-system, since it is quicker to ask a (highly specific) question repeatedly than to search for a similar question in a database, which is probably too generic anyway.

However the communities of BP and Shell both communicate mainly via the communication tools, which, in some communities appeared to happen quite frequent. In the communities of Unilever communication appeared to be mostly restricted to the face-to-face meetings held once or twice a year, with little communication between these meetings. The AtosOrigin Expertise Group members communicate practically exclusively through face to face meetings.

Providing much ICT support and expecting that the group will function optimally *because* this support is available may be detrimental to the success of the group (see Delft Cluster). This is valid for teams, but may even be more valid for communities, since there is not a strong common task that pushes for interaction. Providing a community with few or many ICT tools may therefore be a real dilemma.

Another dilemma concerning ICT is that between customisation and uniformisation. The application offered to a group should match the needs of this group. So in the most ideal case the community decides itself which ICT application it wants to use and how to customise that tool. On the other hand, however, it is stimulated to use uniform platforms and applications in order to make it possible to communicate and share applications between different locations.

CONCLUSIONS

In their search for a viable knowledge management strategy, but also inspired by concepts of organisational learning, many organisations now opt for a 'personalisation strategy'. They expect a community of practice to accomplish what knowledge technology has failed to do. Five types of CoPs of practice where identified above, of which two – the 'formal expert community' and the 'informal network community' – seem to be most frequently found in companies. The purpose pursued and the level of formality appear to be major distinguishing dimensions between the various types of CoPs. These and other factors imply several dilemmas for communities. They can be concentrated in the following questions:

- How important is the role of size, cohesion and trust? Does knowledge sharing require a small, cohesive group, where members know and trust each other, or can effective knowledge sharing and creation also take place in large loosely coupled groups?
- How to resolve the dilemma of formality versus spontaneity? Knowledge sharing and creation seems to bloom in informal, bottom up settings. On the other hand, continuity may require formal structures, with carefully chosen members, leaders and procedures.
- How can ICT tools support the community processes optimally? Practical examples seem to suggest quite divergent ideas: some communities seem to be hindered by sophisticated tools, others seem to thrive on them.

But most important is the question how to organise the interaction of Communities of Practice and of knowledge systems in such a way that both individual and organisational learning are optimal.

REFERENCES

Allee, V. (2000). *Knowledge Networks and Communities of Practice.*
     http://www.odnetwork.org/odponline/vol32n4/knowledgenets.html Dec. 2000
Anderson, N. and M. A. West(1998). Measuring climate for work group innovation: development and validation of the team climate inventory. *Journal of Organizational Behaviour,* **19**, 235-258.
Andriessen, J. H. E., M. Soekijad, M. Huis in 't Veld, and J. Poot (2001). *Dynamics of Knowledge Sharing Communities.* Report for the Telematica Instituut, Hengelo.
Andriessen, J. H. E. (2002). *Working with groupware. Understanding and evaluating interaction technology.* Springer Verlag: London.
Argyris C. and D. A. Schön (1996). *Organizational learning II; theory, method, and practice.* Addison-Wesley, Amsterdam.
Argyris, C. (1992). *On organisational learning.* Blackwell, Cambridge/Oxford, UK
Boomen, M., van den, (2000). *Leven op het Net. De sociale betekenis van virtuele gemeenschappen.* Instituut voor Publiek en Politiek, Amsterdam.
Brandenburger, A. M., and B. J. Nalebuff (1996). *Co-opetition.* Currency-Doubleday, New York.

Brown, J. S. and P. Duguid (1991). Organizational learning and communities-of-practice. Towards a unified view of working, learning, and innovation. *Organization Science*, **2**, 40-57.

Bruin, H. d., and C. d. Neree tot Babberich (2000). *Competing values in knowledge management*. Lemma Publisher, Utrecht:

Campion, M.A., E. Papper. and G. Medsger (1996). Relations between work team characteristics and effectiveness: A replication and extension. *Personnel Psychology*, **49**, 429-445.

Collison, C. (1999). Connecting the new organisation. How BP Amoco encourages post-merger collaboration, *Knowledge Management Review*, **7**, 12-16

Guzzo, R. and M. Dickson (1996). Teams in organizations: recent research on performance and effectiveness. *Annual Review of Psychology*, **47**, 307-338.

Hackman, J. R. (1987). The design of workteams. In: *Handbook of organizational behaviour* (J. W. Lorsch, ed.), pp. 315-342. Prentice Hall, Englewood Cliffs, NJ.

Hansen, M. T., N. Nohria and T. Tierney (1999). What's your strategy for managing knowledge. *Harvard Business Review*, **77**, 106-116.

Hildreth, P., C. Kimble and P. Wright (2000). Communities of Practice in the distributed international environment, *Journal of Knowledge Management*, **4**, 27-38.

Huysman, M. and D. de Wit (2002). *Knowledge sharing in practice*. Kluwer, Dordrecht, The Netherlands.

Klein Woolthuis, R. A. (1999). *Sleeping with the enemy: Trust, dependence and contracts in interorganisational relationships*. Dissertation. Febodruk, Enschede, The Netherlands

Kramer, R. M. (1999). Trust and distrust in organisations: Emerging perspectives, enduring questions. *Annual Review of Psychology*, **50**, 569-598.

Lave, J. and E. Wenger (1991). *Situated learning. Legitimate peripheral participation*. University Press, Cambridge.

McDermott, R. (2001). Knowing in community: 10 critical success factors in building communities of practice, http://www.co-i-l.com/coil/knowledge-garden/cop/knowing.shtml

McGrath, J. E. (1984). *Groups: interaction and performance*. Prentice-Hall, Englewood Cliffs, NJ.

Nonaka, I. and H. Takeuchi (1995). *The knowledge creating company*. Oxford University Press, New York.

Poole, M.S. and G. DeSanctis (1990). Understanding the use of group decision support systems: The theory of adaptive structuration. In: *Organizations and communication technology*, (J. Fulk and C. Steinfield, eds.), pp. 173-193. Sage, Newbury Park, Ca.

Sundstrom, E., K. P. DeMeuse and D. Futrell (1990). Work teams. Applications and effectiveness. *American psychologist*, **45**, 120-133

Weggeman, M. (1997). *Kennismanagement: Inrichting en besturing van kennisintensieve organisaties*. Scriptum, Schiedam, The Netherlands

Wenger, E. and W. M, Snyder (2000). Communities of Practice: The Organizational Frontier. *Harvard Business Review*, **78**, 139-146

12

KNOWLEDGE MANAGEMENT IN FOUR SERVICE ORIENTED COMPANIES

Steven Dhond, Veronique Verbruggen, Guurtje van Sloten and Tobias Kwakkelstein[1]

INTRODUCTION

In chapter 4, a hypothesis was developed in which knowledge transfer systems were related to organisational type. This hypothesis was developed to help explain why in the practice of knowledge management, implemented knowledge management systems are little used (Damodaran and Olphert, 2000) or knowledge management systems rarely help to support the goals companies have (Strikwerda, 2000). The goal of this chapter is to look into knowledge management systems and the consequences for quality of work. If knowledge is power, what does knowledge management signify for the power relations in the company? How do companies deal with knowledge management in their day-to-day doings? We will try to answer these questions by looking at four case studies in the service sector which have recently implemented knowledge management systems (Verbruggen and Wiezer, 2000). In the last chapter of this chapter we will discuss these results and try to point out what these conclusions mean for knowledge management implementations.

---

[1] TNO Work & Employment, Hoofddorp, The Netherlands

CASES

Our research was oriented at four service oriented companies. In earlier research oriented at those companies which proclaimed themselves as knowledge management proponents, we uncovered that these companies were not so much oriented at knowledge management, but rather at competence management (Wiezer and Mossink, 1999). These companies were mainly production companies in which production disturbances were rather rare. Knowledge development and innovation were tasks of central staff and research departments. Such machine bureaucracies were only interested in the creation and control of explicit knowledge. Therefore we have oriented ourselves at service companies which were clearly dependent on (operational) knowledge creation and development to survive in their market. Four companies in different service sectors were selected to have sufficient organisational differentiation. The four companies were:

- IT-service: this is a service organisation which is active in the production and service of hard and software systems. This company is organised around market projects. Professionals are responsible for these projects. Next to these professionals, a central R&D-division controls product development and innovation.
- Insurance: this insurance company is quite centralised and concentrated. Central staff departments and central management have a great influence on all decisions taken. All operations are located within one building, though the organisation orients itself at the national market.
- Consulting: this is a non-profit consulting organisation. Consulting is a project based company. Subsidy systems guide the professionals in selecting new projects. In this organisation, the central management is engaged in an effort to acquire more control on its operations.
- Nature: this is a national branch of a world wide operating nature protection organisation. The national branches have the edge on a weak international management. National managers control the funds in the organisation. International programs are under construction, but these programs do not signify a change in current practices.

These four companies possess strongly deviating organisational forms and knowledge management practices. In each of the companies a series of interviews was conducted with management and worker representatives. Interview reports and final reports were given back to the companies for comment and correction. This material has been used for further analysis (Verbruggen and Wiezer, 2000).

KNOWLEDGE MANAGEMENT IN THE FOUR COMPANIES

IT-service

At IT-service, the knowledge management system is called 'knowledge mastering'. This system resembles a project management methodology. By means of special trainings and development programs, consultants are educated. The development programs have as objective to create more free time and to consolidate experience. The consultants in the training programs have to take great care to optimise their knowledge and systems, and to consolidate their knowledge. Knowledge mastering is completely controlled by the consultants. They are supported by tools and persons, but it is the consultant who gives content to the knowledge mastering. The control system is built on the concept of a learning cycle and wants to secure an optimal input of the existing knowledge. The control system starts with the projects in which the consultants participates. Information such as project inputs, project development, project execution, output and evaluation are included into the system.

The organisation of knowledge management, the development and consolidation of the experiences are the responsibility of the working group 'fields of interest'. Each field of interest is a separate market (or 'opportunity') within which IT-service is active. Such fields of interests have a global orientation. Each field of interest is divided into separate competence areas. All fields of interest can be categorised into either technological or business related fields. Certain competence areas are covered by several consultants and others are still in full development.

The main underlying idea is that of the 'learning community' structure. The hypothesis is that the best knowledge is experience based. At local and regional level, meetings are organised by consultants active in a same competence area. This project based exchange of experiences is supplemented with tools which are delivered by a utility workgroup. Starting from these experience exchanges (tacit-tacit), a translation is made from 'best practice' to more structured knowledge (tacit-explicit). This structured knowledge is called the 'Structural intellectual capital' or SIC. This SIC can vary from the offering of new services, catalogues in databases or new approaches and methodologies.

This learning community structure is presided by a 'spokesperson' who has a worldwide perspective and who has several pupils under his guidance. These pupils are the regional spokespersons which themselves lead the 'local learning community'. The experiences at the local and regional level are exchanged and further enhanced. By means of a spiral movement of feedback processes, these experiences are further transferred to the 'global' spokesperson. This person has the task to structure these experiences and to put them into the Knowledge-Net and the Knowledge helpdesk. The K-Net gives an oversight of the best practice cases at the world level in the form of case evaluations, new products or process development sketches. In this way, the K-Net manages and consolidates knowledge in the organisation. The K-Net is called a knowledge broker. The K-Net shows how to find answers to questions. The K-helpdesk consists of several persons who can give support in the handling of certain problems and who can help to translate problems into best practice cases.

Insurance

At Insurance, innovation is achieved by the combination of existing products to new products. The market power of Insurance depends greatly on the speed with which it brings new products to the market. To have a stronger influence on this innovation process, Insurance has decided to strengthen the co-operation between specialists and the insurance intermediaries. The organisational structure has changed from a product orientation (fire insurance, life insurance, etc.) to a product-market orientation (private persons, companies, employee benefits). The insurance intermediaries are now 'connected' to a regular correspondent in the organisation, which is a change from the past where they only had correspondents for each type of product. This market solution breaks up the insurance product specialists into different market groups. The main risk Insurance incurs with this new structure is that the speciality knowledge and insights in law developments is dispersed throughout the organisation and in this way dissipates. Knowledge management is used as a means to secure this central knowledge and skills for the organisation (tacit-explicit). The main means are a *central knowledge database,* which contains up-to-date information on all products, and the *knowledge coordinators.* The knowledge coordinators work within the different product areas and are responsible to detect knowledge shortages and learning needs. They also decide on the learning method which is required to fill the knowledge gaps within the organisation. The knowledge coordinators are supported by specialists, supplying the day-to-day answers to the questions put forward. The specialists also look after the quality of the information in the knowledge database. A set of meetings and communication structures completes the set of knowledge measures. The meetings are needed to detect possible flaws in the knowledge production in the company.

Consulting

The main task of Consulting is to stimulate technological and product innovation in small and medium sized companies (SME's). A Consulting advisor has to make entrepreneurs more vigilant for external developments and has to help him to react to these developments. Consulting is also responsible for the knowledge transfer from the knowledge providers (universities, research institutes) to these SME's. A first reason to start knowledge management is the pressure from the financing ministry to become more accountable for the projects and the project results. This ministry requires a detailed reporting system on these projects. A second reason for knowledge management is the will of the management to achieve a greater customer orientedness, to better facilitate the knowledge transfer between knowledge providers and to support the professionalisation of the innovation advisors. In fact, there are two knowledge management projects: an external and an internal knowledge management project.

External knowledge management. This project has started because of the request of the financing ministry. Consulting is now trying to help SME's to develop their own knowledge management system. The methodology developed for this goal is called 'Knowledge Analysis Methodology'. Next to this consulting track, there are the Knowledge Platforms. These platforms are managed by a country wide team which consists of several regional directors and consultants. In such a platform, attention is directed towards areas which are of importance for SME's in the future. The platforms function as an interface between knowledge providers and the SME's. The following platforms are currently active: agro technology, product development and innovation, materials technology, intelligent products and processes, sustainable development, ICT, innovation management and marketing, HRM and co-operation.

Internal knowledge management. The internal knowledge management project is built around the 'knowledge sharing' initiative. The management of consulting tries to create an environment in which sharing of ideas and contacts between consultants is enhanced. At this moment, knowledge creation is still a too individual process. In this new initiative, consultants are supported by the ICT-department and the New Technologies group. ICT helps the workers via the helpdesk and functional support at the workplace. The main application which is meant for knowledge sharing is Lotus Notes. Lotus Notes consists of three main databases: contacts, projects and scouting. But there are still too much other databases outside Lotus Notes which are difficult to access. The main system is still not interactive so that interest from consultants in sharing ideas fades quite quickly. The main task the ICT-department has got, is to rationalise these databases and to develop more interactive means to use these databases.

New Technologies is a separate R&D-department within Consulting. This department tries to identify new developments within Consulting. Whenever necessary, consultants from different regional offices are brought together to participate in these new initiatives. New Technologies works together with the central ICT-department whenever ICT-support is needed for these initiatives.

Nature

Nature has taken up knowledge management as a means to cope with the complexity of nature conservation. Nature conservation is now oriented at implicating whole regions into initiatives and at finding solutions which are broader than biological conservation. In these new initiatives, a whole range of different stakeholders and perspectives must be integrated. For such a strategy, it is necessary to have the field workers communicate more and more with other parts of the organisation. Experience sharing is becoming the main road for the organisation to deal with this new strategy. Knowledge management was started by means of a pilot project in one of the regional offices of the organisation. The project is called the 'Action Network'. We will show what the main elements of this knowledge management project are. The main elements in the new conservation strategy are ecoregions and the concept of 'magnification'. In the past, Nature was mainly oriented at conserving certain species. Now it is clear that species can't survive if their complete habitat is destroyed. Therefore, Nature wants to have projects which have the scope of ecoregions, which can stretch out to different countries. Magnification is the way Nature tries to engage local groups into such an ecoregion project. By using her 'golden assets' (specialised knowledge and expertise, worldwide network, reputation), Nature tries to develop partnerships between local and national authorities, indigenous population, industry and international organisations to combat the disappearance of ecological valuable areas.

The Action Network has been started by a small team in one of regional offices of Nature. The team members are the change agents who try to motivate the organisation to use and adapt the knowledge management instruments developed during the project. The change agents use a mix different knowledge areas to achieve their goals. The main elements of the knowledge management project are dealt with successively.

Programme managers: Ecoregions are not managed by national representatives, but by programme managers who possess a great amount of autonomy to achieve their goals. These new managers are also responsible for publicity around their programmes. One of the means used to achieve such publicity, is the 'living document'. These documents give information to fundraisers about the content of the programmes and about the execution of the programmes. These documents have been made accessible to a broad public via the internet.

Communication managers: Next to the programme managers, there are several communication managers. These are ICT-educated managers who are responsible for the connection between the programmes and the fundraisers. They mainly use the intranet to achieve this knowledge exchange. Communication managers meet quite regularly with programme managers. These meetings are meant to make the programme managers more susceptible to the use of the intranet.

The Intranet: the Action Net: Internal and external communication has been improved by the installation of an intranet, called the Action Net. Local offices are invited to participate in this intranet and are themselves responsible for the content on the intranet. The intranet contains several knowledge areas such as the possibility to collect the important experiences (lessons learnt), question-answer databases, best practices and a 'who is who?'. A separate element in this intranet is a discussion forum which is limited to the programme managers and communication managers. The intranet also gives an overview of the competencies and specialties of each of the workers of Nature. The use of the intranet was rather limited in the start-up phase. For this reason, the change agents organised several events and face-to-face meetings to improve the interaction between workers. The intranet is also used as a means for the reporting out in the field. Programme managers can report directly in writing, in picture or with video materials.

College for Conservation Leadership: A separate initiative is a central schooling system to help develop high potentials in the organisation and to improve competencies within the organisation. This initiative consists of classical schooling programmes, but also of means for remote learning and video conferencing. Trainees are supported by a whole set of coaches, companions and experts.

Action Room: The intranet is not sufficient to support the networks of communication within Nature. For that reason, a centralised Action Room was created. This Action Room is independent from the ICT department, and is responsible for helping the organisation in realising better communication procedures. Mainly, these means are via databases and/or ICT. The Action Room functions as a helpdesk for the organisation. In the future, this Action Room should be the centre of the organisation to combat certain 'catastrophic events'. The different technological means (for example satellite communication) make this instrument most suited for this task.

IT-service

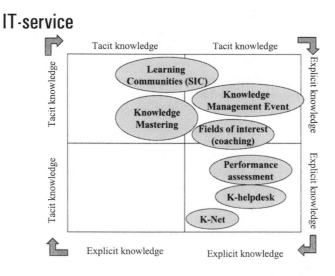

Insurance

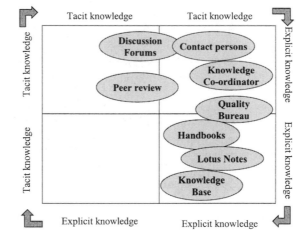

Consulting

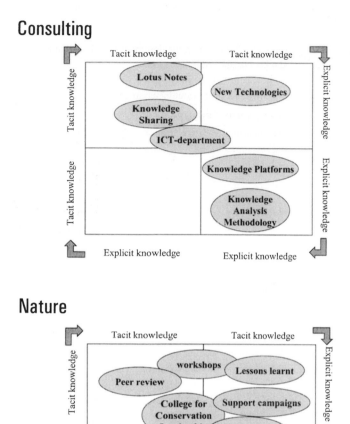

Nature

Figure 12.1: Dominant knowledge transfer mechanisms for the four case studies

ANALYSIS

Table 12.1 gives background information on the dominant knowledge transfer mechanisms in the four companies and other background information.

Table 12.1: Knowledge management in the four different companies.

	IT-SERVICE	**INSURANCE**	**CONSULTING**	**NATURE**
Dominant knowledge management transfer mechanism	Externalisation and socialisation	Externalisation and socialisation	Socialisation	Socialisation
Motives knowledge management	External pressure	External pressure	Internal pressure	External pressure

The table shows that two different models of knowledge transfer are used. IT-service and Insurance give priority to externalisation as knowledge transfer model. These companies systematically try to find which new solutions are used by workers. These solutions are either included into current procedures or are excluded by reverting to older procedures. Nature and Consulting follow another path. In their knowledge management model, they give priority to socialisation. These companies try to motivate workers to give on their newly acquired knowledge to their fellow workers. The externalisation method, although present, is not so well developed.

The motives to start knowledge management in the four companies are not always the same. For IT-service, Insurance and Nature, the central motive to develop knowledge management is because of changes in the company strategy. These strategic changes have been accompanied by organisational changes. External pressures are the main reason for these strategic reorientations. For Consulting, internal pressures were the prime factor to start knowledge management. The sudden initiative by some workers to start knowledge management were the main reason for the management to bring the initiative at the company level. Of course, the pressures of the major financing ministry spurred the centralisation tendency in knowledge management. In this company, management wanted to develop its own knowledge management model, starting from scratch. Although management recognised the merit of the worker initiative, it was mainly interested in a model which was under its control form the start.

The knowledge management strategies are not unidirectional. In none of the cases, there is a clear-cut path for the development of knowledge management. Each of the programs has several goals. Transparency of operations is the prime goal, but in one of the companies knowledge management is also used as a means to make the company more attractive on the labour market.

KNOWLEDGE MANAGEMENT AND QUALITY OF WORK

Table 12.2 presents our findings for the influence of knowledge management on quality of work.

Table 12: Influence of knowledge management on quality of work in the four companies.

Influence on:	IT-SERVICE	INSURANCE	CONSULTING	NATURE
▪ complexity operational tasks	+	-/+	+	+
▪ control possibilities	+	-	+	+
▪ tacit knowledge developed?	-	-	-	+
▪ participation of workers in the total process of knowledge management	+/-	-	+	+
▪ labour market position of operational workers	+	+/-	+	+/-

From table 12.2 we can learn that knowledge management doesn't mean a worsening of quality of work. Only Insurance shows a deterioration of control possibilities of workers. We see that workers in this company have no role in the shaping of knowledge management. We can also see that knowledge development possibilities for operational workers are less. These main results cover the fact that these consequences might be different for different groups of workers in the companies. This is the case for instance in Insurance and Nature where we can see that different jobs have seen several tasks dissipate with the implementation of knowledge management. Mainly middle management is a 'victim' in this sense that they experience a loss of power in comparison to workers they supervise. Newly created jobs are the winners of the new situation because important tasks are transferred to these jobs. Knowledge management does not have a unidirectional influence on quality of work. It is of importance to look at the organisational context within which knowledge management is implemented and to the goals knowledge management has been given. Another remark that can be made is that the influence of knowledge management on quality of work may not be overstated. Other company measures may have more impact on quality of work than knowledge management. For most workers, knowledge management brings a strengthening of the labour market position. In this sense, it is understandable that workers in general are positive towards the implementation of knowledge management.

ORGANISATIONAL DIFFERENCES AND KNOWLEDGE MANAGEMENT

Our last research question is if knowledge management is influenced by the organisational context within which it is embedded. Table 12.3 contain this result. We have looked at the way these companies have organised the planning, administration and further development of knowledge management. We have also compared the current organisational type with the 'intended' organisational model. The intended model is the one we could discern in our interviews with these companies.

Table 12.3: Knowledge management in the four different companies.

	IT-SERVICE	INSURANCE	CONSULTING	NATURE
Control model knowledge management	Line	Quality management	Combination line and staff	Combination line and staff
Current organisation	Professional bureaucracy	Machine bureaucracy	Professional bureaucracy	Divisional form
Intention	Professional bureaucracy	Machine bureaucracy	Simple structure	Machine bureaucracy

The final table shows that the four companies use different control strategies for knowledge management. IT-service has built its knowledge management into the existing management structure. Managers and operational workers have a direct control on the knowledge management process. There is no significant role for staff representatives. At Insurance, knowledge management has been concentrated into the tasks of the quality management staff. The main reason for this company to do this, is that standardisation of knowledge is important for the company to guarantee the required security of its services to customers. One wouldn't want insurance contracts ridden with quality defects. Consulting and Nature have hybrid knowledge management models. The current managements but also operational workers are responsible for planning, administration and development of knowledge management. These companies show the strengthening grip of the central staff services on the knowledge development process.

With this information, we can look back at the organisational contexts within which these knowledge management programmes are introduced. IT-service and Consulting are currently working according to the professional bureaucracy type from Mintzberg (1983). IT-service doesn't show a tendency to change this model. The stress put on externalisation as knowledge transfer model is in contradiction with our hypothesis in table 4.1 in chapter 4. Apparently, although much of the organisational power resides in the professionals, central management wants to gain more control on knowledge generation by means of knowledge management. Management does however recognise the fact that the professionals are the core of the company. In Consulting on the other hand, we can see that management wants to achieve a greater power on the organisation. The hope of management is to change over to a simple structure. It is clear that in this organisation, a power struggle is going on between management and operational workers. Knowledge management has recently become subject of this power struggle. In Insurance, the power of the staff services remains untouched. The organisation resembles a machine bureaucracy and management has no intention to change this fact. Knowledge management has been shaped in such a way that the staff service 'quality management' is responsible. From this perspective it is understandable that externalisation is the main thrust in knowledge management. Nature is shaped according to the divisional form. The new knowledge management is introduced to give more power to staff services. From this development, we can deduce that the intent of the weak central management is to change over to a machine bureaucracy type of organisation. The current model of knowledge management is still oriented at socialisation. We may expect this model to change over to a more suited model.

DISCUSSION

In this chapter we have analysed the knowledge management schemes of four service companies in The Netherlands which rely on knowledge of their workers to compete in their markets. The main focus was the influence of these knowledge management schemes on work practices in these companies. We wanted to explain differences in schemes by looking at the organisational type used by the companies. These organisational types differ from one another in the power centre the company has. Our hypothesis was that there is a 'fit' between organisational type (Mintzberg-types) and knowledge management model (Nonaka-typology, Nonaka and Takeuchi, 1996).

Our conclusion is that companies use different approaches to knowledge management. We have seen that the stress was on two dominant approaches of knowledge management: an 'externalisation approach' on the one hand, and a 'socialisation approach' on the other hand. In the first strategy, great attention is given to making all tacit knowledge explicit. In the second strategy, tacit knowledge is passed along between workers. Our hypothesis about a fit between organisational type and knowledge management model was only partly validated. We found that organisations are in transition and power struggles are an ongoing process. This makes it clear that the intention of knowledge management may not always fit the organisational type. In our investigation, we found that IT-service is organised according to a machine bureaucracy, but knowledge management supports the power of central management. It is our expectation however that this knowledge management model will experience a lot of tensions. Since operational workers have the main power in this organisation, central management will not always be able to achieve its goals with knowledge management. The fit between organisational type and knowledge management works both ways.

Working situations appeared to profit from knowledge management in this sense that work was becoming more complex and responsibilities were decentralised. Only in Insurance, which resembles a classic tayloristic organisation, we can find that quality of work is poor. These results are not always the same for all groups in the companies. Middle management is a 'victim' in this sense that they experience a loss of power in comparison to workers they supervise.

Knowledge management approaches are always oriented at core tasks of companies. But the control strategies tended to differ between the companies: two approaches seemed to emerge, a line approach and a functional approach to knowledge management. All in all, it seems that knowledge management, although important for these companies, didn't shock the major operations of the companies. The impact of knowledge management is dependent on other policies such quality management, personnel policy etc..

What do these results signify for further research on knowledge management? We see two major implications. A first implication is for the implementation of knowledge management in companies. Knowledge management is clearly not a neutral activity. Knowledge management obscures the fact that companies are ridden with power struggles. Some of these struggles are open, most of them are hidden. One must not be mistaken that the introduction of knowledge management will be an easy thing to do. Win-win situations are rare. The concept of knowledge management covers the fact that many managers are reorganising companies and are not so much interested in exploiting hidden knowledge. It must at the same time be clear that the implementation of knowledge management must take into account what the organisational type is. If knowledge management is to be successful, one has to recognise the interests of the most powerful group in the company. Also, technology is certainly not a neutral instrument in this process. Technology can be used to brake the power of certain power groups in the companies (Majchrzak and Borys, 1998). At the same time, technology may have unexpected uses which could frustrate the original intention of a knowledge management introduction (Ciborra, 1996).

A second implication is for the way future research into knowledge management should be conducted. We have conducted four case studies. Our conclusions can only be tentative for this subject. It is necessary to enlarge the field of research to more cases and even to surveys. This last possibility is not easy, because of the nature of the research. Power struggles are very difficult to measure with questionnaires. The models used in this chapter may certainly be of use in conducting such broader research.

REFERENCES

Ciborra, C. (eds) (1996). *Groupware and teamwork. Invisible aid or technical hindrance?* Wiley, New York.

Damadoran, L. and W. Olphert (2000). Barriers and facilitators to the use of knowledge management systems. *Behaviour & Information Technology*, **19**, (6), 405-413.

Majchrzak, A. and B. Borys (1998). Computer-aided technology and work: Moving the field forward. *International Review of Industrial and Organizational Psychology*, **13**, 305-354.

Mintzberg, H. (1983). *Structures in fives: designing effective organizations.* Prentice Hall, Englewood Cliffs.

Nonaka, I. and H. Takeuchi (1996). *De kenniscreërende onderneming*, Scriptum, Schiedam.

Strikwerda, J. (2000). De beperkte visie van de consultant. *Management Consultant*, **5**, 46-49.

Vaas, S. and S. Dhondt (eds) (1995). *De WEBA-analyse.* Samsom, Alphen aan den Rijn.

Verbruggen, V. and N. M. Wiezer (2000). *De praktijk van kennismanagement bij vier dienst-verlenende bedrijven. Casestudies.* TNO Arbeid, Hoofddorp.

Wiezer, N. and J. Mossink (1999). *Sociaal beleid en strategie.* TNO Arbeid, Hoofddorp.

13

PRACTICES OF MANAGING KNOWLEDGE SHARING

Marleen Huysman[1]

INTRODUCTION

This chapter will present case study findings on practices of managing knowledge sharing. The research includes ten structured knowledge-sharing practices within ten large companies with more than thousand people working. In all these companies, top management encouraged the initiatives. All initiatives were supported by information communication technology (ICT). Along with the types of knowledge sharing that we studied, these ICT applications ranged from knowledge bases to electronic communities.

As the research concentrated on actual experiences and possible problems, we needed to use a different selection of cases than selecting knowledge management initiatives only. Because we only included initiatives that had already been running a substantial period, knowledge management practices were excluded from which we could not extract empirical lessons. This meant a serious shortcut of potentially interesting case studies. In fact, most knowledge management initiatives that we at first considered as being interesting enough to include in the study appeared to be still in a conceptual stage. As a result, we decided to broaden the range of potential cases by also including practices of managing knowledge sharing that had not (yet) received the label 'knowledge management'. Consequently, we obtained a rather pragmatic

---

[1] Vrije Universiteit Amsterdam, The Netherlands

orientation to knowledge management. In short, *knowledge management* is perceived here as *organisational practices that facilitate and structure knowledge sharing among knowledge workers*. With successful knowledge management we refer to practices of knowledge sharing that have become embedded in the ongoing work processes of an organisation. In other words, we perceive the success of knowledge management to be related to the degree in which sharing knowledge has become a taken for granted part of the routine practices within the organisation.

Because of the ambiguity within the literature in general, we opted for an exploratory case study research. To do so, we asked the following research questions: who's knowledge is managed, what knowledge is managed, when is knowledge managed, why is knowledge managed, how is knowledge managed and where is knowledge managed? We believe these questions are by far the most reasonable ones that can be conceived of when trying to explore a concept in practice.

The theoretical model introduced in chapter 3, will be used to analyse knowledge sharing in practice. Consequently, we make a distinction between practices to manage knowledge acquisition, knowledge exchange and re-use, and knowledge creation. (Table 13.1 presents the primary purpose of managing knowledge sharing related to the type of learning, the type of ICT application they used to support the knowledge sharing activities and the various companies where we studied knowledge sharing initiatives. Because of space limitations, we refer the interested reader who would like to learn more about the individual cases, to Huysman and De Wit (2002).

Table 13.1: Various types of knowledge sharing in practice

Types of knowledge-sharing	Knowledge acquisition	Knowledge exchange and re-use	Knowledge creation
Learning from:	Organisational knowledge	Individual knowledge	Community knowledge
Main purpose	Store dispersed collective knowledge to enhance individual learning	Prevent occurrence of knowledge gaps and redundancy	Combining knowledge to create new ideas and insights
(ICT) support	Knowledge bases	Knowledge bases and networks	Networks
Companies studied	▪ Postbank's Call Center ▪ National Netherlands ▪ Railways	▪ Cap Gemini ▪ IBM ▪ ING Barings ▪ Airport	▪ Unilever R&D ▪ Stork ▪ Ministry of Housing

KNOWLEDGE SHARING IN PRACTICE

Knowledge acquisition

The insurance company the 'Postbank', 'National Netherlands' (NN), and the Netherlands Railways, offered insights about their experiences with providing organisational members access to organisational knowledge. Table 13.2 provides a summary of the findings related to knowledge retrieval at the three companies.

Table 13.2: Knowledge acquisition initiatives

Learning Process	Postbank	NN	Railways
Type of process	Organisation as knowledge owner. Domain knowledge for training on the job; match practice with learning.	Storing knowledge. Integration domain knowledge, individual learning plus personal networks.	Support of mobile personal on the train
Support of learning process	Knowledge bases and interactive learning environments	Knowledge bases. Structured physical networks.	Knowledge bases
Purpose	Knowledge storing, mind mapping and actualizing to increase client satisfaction, to support and socialize operators	Retain knowledge because of change from product to market oriented organisational forms.	Facilitate job performance
ICT	Infobase	Knowledge base	NS rail pocket (mobile knowledge base)
Role of ICT	Essential, regular use	Essential, regular use	Essential, regular use.
Some experiences	No feedback, speeding up task execution.	Friction between hierarchy and knowledge sharing responsibilities	Dependency on knowledge bases, no feedback, no monitoring
Type of worker	Salespersonal at call center	Insurance and knowledge worker	Conductor

In the case of the Postbank, sales personnel at the front office use organisational knowledge that was stored in a knowledge base. This knowledge base was used not only to support client interaction but also to support the socialization and training of the call center operators. In the case of the NN, insurance employees make use of organisational knowledge that is stored in a knowledge base. In addition, employees refer to a structured personal network of insurance experts to learn knowledge about the operational processes of the organisation. At the railways, train conductors use a mobile knowledge base 'the railpocket' to gain knowledge about the operational process.

In all three cases, the purpose of the knowledge sharing initiative was to retrieve organisational knowledge in order to use it for operational processes. Typical to these knowledge-sharing initiatives is that the organisation provides access to explicit formal organisational knowledge in order for individual members to learn from and to transfer it into implicit knowledge. This learning process can be described as 'internalisation': knowledge transfer from organisation to individuals.

All three companies intensively used knowledge databases. The insurance company and the bank used a knowledge database in their sales and service operation. The bank had created this database through mind mapping techniques and transferring existing documents. Both organisations felt a need to create a knowledge database when they transformed from a product driven organisation to a market driven organisation. The transformation implied that people had to operate as generalist instead of specialists. In order to maintain an appropriate level of service, employees needed to be supported. In both cases the database was linked to a learning system. New employees could use the database as tool to quickly become a generalist. The insurance company also created a network of people. Knowledge co-ordinator and knowledge specialist were roles needed to secure the development of new knowledge, and also provide a fall back position for questions of employees.

The Netherlands Railways provided an interesting example. In this case an employee developed a mobile application to overcome the weight of carrying around travel schedules and handbooks. Senior management quickly took up the idea. The mobile application, the so-called 'Railpocket' contains organisational knowledge and gives employees room to fill out their administration and day-to-day experiences.

In all three cases it was possible to discern some problems of managing and supporting knowledge retrieval. We will discuss these in the next paragraph.

Knowledge re-use

Our research provided four illustrations of structuring knowledge sharing initiatives with the purpose to exchange and re-use knowledge. These companies were IBM, Cap Gemini, Airport Schiphol and ING Barings (see table 13.3).

Table 13.3: Knowledge re-use and exchange initiatives

Learning process	Schiphol	ING	Cap Gemini	IBM
Type of process	Storage of knowledge such as about personal networks	Making knowledge within various countries accessible	Re-use	Re-use
Support of learning process	Knowledge centre	Network of countries	Informal networks and electronic networks	Electronic networks
Purpose	Capture personal knowledge and identify networks	Make country information accessible	Make knowledge accessible and increase efficiency	Make knowledge accessible Standardization, ready made solutions
ICT	Network	Intranet	Intranet	Intranet
Role of ICT	Limited	Important	Important	Essential
Some experiences	Reassure cooperation	Problems in mobilizing geographically dispersed people	Differences in bottom up and top down initiative, time limitations	Part of transformation, problems in time and recognition
Type of worker	Policy.	Corporate finance, policy	Consultants	Routine consultants

The purpose of these initiatives is to let people learn from each other through re-using individual knowledge. Both social and technical networks are used in order to facilitate this learning process.

In the case of Schiphol, a knowledge centre was initiated that captured personal knowledge and identified existing networks in order to support the knowledge work of policymakers at the Airport. At ING Barings, an intranet was introduced to support knowledge exchange between the different countries. At Cap Gemini, consultants use both informal personal networks and electronic networks (Cap Com and the Galaxy), to enable re-use of knowledge. Re-use of knowledge is also the main purpose at IBM to install an intranet. Consultants at IBM use this codified knowledge especially in case of standardization and ready made solutions.

Re-use of knowledge created by knowledge workers is a common argument to start a knowledge management system or an Intranet. Especially consultant firms see re-use of knowledge as an important strategy. Organisational practice (using inexperienced people providing fit-to-the-situation solutions) and customer pressure create this need for re-use.

An important driver for these organisations to introduce knowledge management is the high turnover of personnel. Management needs to look after their knowledge base, with typically 'best practices' as an application. Another motivator is customer driven. Customers, especially clients of knowledge intensive organisations such as the ones discussed here, ask for knowledge that has been proven valuable elsewhere. They do not seek for new solutions but rather have a cost-effective solution that has been used already. The different needs and interests also lead to one of the fundamental problems: the difficulty for knowledge workers or professionals to engage in knowledge sharing. Most knowledge workers will find it hard to express what is meaningful in their work. Also, professionals are more often focused on developing their own solutions instead of using other people's ideas (Weggeman, 1997). Characteristic of these organisations is that professionals develop their own networks to obtain the knowledge and information they need. High workload and high time pressures often prevent knowledge workers from externalising their knowledge in best practices unless a social aspect is part of the process.

In a later section we discuss which problems these four companies experienced in managing knowledge re-use.

Knowledge creation

Although many organisations start with knowledge capturing, often an ultimate goal is creating new knowledge. In an organisation with fixed routines and procedures, this need for generative learning may be less than in an R&D environment. But still, creating new ideas and insights through sharing knowledge is on many business objectives list. We studied knowledge creation in three different organisations: Unilever, Stork (a high tech multinational company) and the Ministry of Housing (see table 13.4). In various ways these organisations made use of communities or groups of individuals with shared knowledge interests.

Table 13.4: Knowledge creation initiatives

Learning process	Stork	Unilever	Ministry of Housing
Type of process	Sharing to get new ideas	Sharing to re-use and develop new knowledge	Sharing to get new input for future policy
Support of learning process	Study groups	Knowledge workshops	Virtual communities
Purpose	New insights	Exchange and development of knowledge	Interactive policy development
ICT	Electronic rapports	Knowledge mapping, Lotus Notes	Digital discussion platforms
Role of ICT	Marginal	Present but not essential	Essential
Some experiences	Structure depends on seniority. Ambiguity about outcomes.	Combination of respect, status and physical encounters stimulates community building	Problems of virtual communities without having clear collective purposes
Type of worker	Experts and managers in high tech	Experts in R&D settings	policy making officials, external stakeholders

Unilever for example gained experience in supporting communities of practice. The company started over five years ago to systematically collect, exchange, create and leverage knowledge because it saw her innovative ideas being copied by her competitors. Using existing knowledge within the company and exchanging this knowledge became one of the strategies to stay abreast. The research department developed this structure as an outcome of so-called 'knowledge workshops'. Unilever now organises workshops in order to bring members together who share their expertise, for example about tomato products, but who are geographically distributed. Initially, these workshops were intended to map this distributed knowledge, to identify knowledge gaps and to store shared knowledge in knowledge databases. Over time the workshops were attributed another more fruitful purpose: the facilitation of communities of interest. Because of the physical encounters between otherwise dispersed people with similar interests, people not only share their knowledge at the moment of the actual meeting, but also tend to get into contact with each other after the workshop. In some cases, these groups become social networks of people who like to explore and develop new ideas. These communities, consisting of geographically dispersed individuals, created fruitful bodies of knowledge that facilitated organisational learning.

The international technology company 'Stork' introduced 'IPI' (integrated process innovation') a structured form of knowledge development through communities. This structure supports different communities enabling them to share knowledge and allow new knowledge to come into existence. Although IPI exists for almost 30 years already, only recently the company sticks the knowledge management label to it. Interestingly, both at Unilever and at Stork little use of ICT was made. Their focus is on connecting people and creating organisational networks rather than technical one. The Ministry has 'electronic communities'. It uses the Internet so that communities focussing on a specific theme can exchange thoughts. The Ministry also incorporates the results of this type of electronic discussion platform in its future policy development.

In the next section we will discuss the experiences with managing knowledge sharing that these and the other seven companies gained over the years.

ANALYSIS: IDENTIFYING TRAPS AND WAYS TO AVOID THEM

We addressed our research material with six exploratory research questions in mind (see table 13.5).

Table 13.5: Six research questions and their dominant biases and related traps.

Research question	Knowledge sharing bias	Knowledge sharing traps
Why is knowledge sharing managed	Control bias	MANAGEMENT TRAP
When is knowledge sharing managed	Opportunity-driven bias	
Who's knowledge sharing is managed	Individual knowledge bias	LOCAL LEARNING TRAP
Where is knowledge sharing managed	Operational level bias	
What knowledge sharing is managed	Codified knowledge bias	ICT-TRAP
How is knowledge sharing managed	Technology driven bias	

While addressing these questions, we observed that all initiatives are or have been biased towards various aspects of managing knowledge sharing. The practices of knowledge management also illustrated that these biases might result in potential traps. Organisations might fall in these traps when being too much focussed on certain aspects while overlooking others.

In the following sections, we will describe the three traps and their related biases separately. Each discussion will be supported with illustrations from practice.

How is knowledge shared and what knowledge is shared: The ICT trap

The focus on *what* knowledge is shared and *how* it is shared helped us to identify a potential risk for falling into a so-called 'ICT-trap'. The ICT trap consists of two different yet related biases, which we will briefly discuss. We encountered this trap when we asked two questions: what knowledge is shared in practice and how is knowledge shared in practice. Addressing the first question revealed that many of the initiatives we studied were biased towards a stock approach to knowledge. Addressing the second question showed that these companies were biased towards a technological driven orientation towards knowledge management. In that case, the underlying assumption is that ICT can support and improve knowledge sharing within an organisation.

Many articles and books on the concept of knowledge management start their discussion with a definition of knowledge. Almost always, the relation is made between two related concepts: data and information. Whereas data are signals and information is signals that make a difference, knowledge is created out of information but is individual specific. In its most extreme definition, knowledge that belongs to individuals cannot be explicated. At the moment we exchange knowledge, the knowledge becomes information to the potential receiver. There is a danger that organisations might treat this externalised knowledge as valuable substitutes for individual knowledge . If so, much valuable knowledge will be overlooked. Because technology supports codifying knowledge, this 'codified knowledge bias' is closely connected to the 'technology driven bias'. Organisations often espouse a technology driven bias because they rely on ICT to make knowledge retrieval possible. There are several problems with focussing too much on codifying knowledge. First, there is the problem of dependency. Organisations may become dependent on their digitised archives with the risk of relying too much on this aspect of knowledge overlooking the value of more fluid and personal knowledge. Secondly, there is the problem of deterioration: knowledge embedded in documents or in expert systems may quickly become outdated. When sharing embedded knowledge is not part of an explicit culture, knowledge databases fall prone to rapid deterioration. This is not a new phenomenon, yet it requires discipline of the 'knowledge worker', which in itself forms another problem of knowledge externalisation. Discipline may be hampered by the pressing agenda. We observed at various organisations the problem knowledge workers have in filling in the knowledge system with past experiences, while already gaining new experiences in a new project or work-environments. Especially in project oriented organisations the pressure to make hours accountable, is high. The obvious solution for management is to create slack in order to enable workers to make their experiences explicit. However, even when creating slack, people will find it hard to make explicit what is truly valuable to the company. We encountered this problem at IBM where engineers were unable to express the valuable learning experiences. On the other hand they are reluctant in using the knowledge documented, because they preferred using their own solution rather than that offered by others. Many authors on knowledge management believe that one of the serious problems with externalising knowledge resides in the unwillingness of knowledge workers to give away their power (e.g. Weggeman, 1997; Wiig, 1999; Davenport, 1997). The argument goes that because knowledge is power, people are selective in externalising their knowledge. We did not come across this argument for rejecting knowledge management. Alternatively, we observed that people seem to be resistant to share their personal knowledge in case going public would increase their vulnerability. In other words, when codifying knowledge would imply opening up individual kept secrets. A final problem is that organisations tend to be focussed more on codified than on situated knowledge. Situated knowledge is knowledge that is not embedded somewhere, neither in manuals nor in the heads of individuals. Instead, individuals interacting with each other create situated knowledge in practice. Situated knowledge is therefore situation rather than individual dependent (Lave and Wenger, 1991).

The technology driven bias is embedded in the conviction that the introduction of technological facilities will improve knowledge sharing amongst people, and harness the organisation against loss of knowledge. Knowledge management is often seen as inherently connected to ICT. For example, the introduction of an Intranet is seen as creating the facility for knowledge exchange – although often in combination with a reward structure meant to encourage people to share their knowledge. Yet, when the technology itself is not fancy enough, or when the use is not adapted to the people working with the technology, people will be driven away, despite rewards or punishments. This will curtail the knowledge management initiative.

We came across several knowledge management initiatives that focused on creating a technological environment, but who where unable to reach the people actually using the system. Most of the Intranets in our study were widely praised but little used by the 'praisors' themselves. One of the awkward effects of the technology trap is that a firm belief exists in improving technology in such a way that earlier barriers are overcome.

Unilever learned its lesson over the past years from falling into the ICT trap. They started out by putting their faith in technology and the opportunities to map expert knowledge in databases. Soon they discovered that creating a network of experts, and facilitating physical encounters opens a large potential for knowledge sharing. The ICT is introduced after the network had become established.

Why is knowledge managed and when is knowledge managed? The management trap

Addressing the question *why* knowledge is managed reveals a bias among managers towards the need to control knowledge. Addressing the question *when* knowledge is managed shows that knowledge sharing is taken up by management and given explicit attention, only when they perceive organisational opportunities to do so. This opportunity-driven bias together with the control bias, was implicitly embedded in many knowledge management initiatives we analysed. Together, these biases increases the chance for organisations to fall into the so called 'management trap'.

One of the most general risks of knowledge management initiatives is that the concept is perceived from a managerial perspective only. Clearly, for managers there are several advantages to manage the knowledge within the organisation. One is that knowledge is often scattered within the organisation. With the emergence of the knowledge economy in which workers gain more and more knowledge specific to their own work process, organisations are in need to make these scattered knowledge domains more transparent. Next and related is the argument that transparency is needed to reduce re-invention of wheels. The ideal is that when everyone knows what everyone knows, people will contact each other to exchange knowledge or to effectively refer customers and clients. Learning from each other has the additional advantage of filling up knowledge gaps that would otherwise exist when people leave the organisation or change positions. The ongoing trend towards globalisation too, calls for the exchange of knowledge among globally dispersed knowledge workers.

For example, ING Barings' Head Office considered it vitally important to build an Intranet, which logged the knowledge from all its branches throughout the world. The idea behind this was that whenever someone from the head offices had to travel to a branch elsewhere, he or she could look up the required knowledge, just before they set off. This could be knowledge for example about the country, the branch, the most important clients or the key prices. This was clearly a situation based on the needs of the head office managers. There was no obvious need for the individual employees at the various local offices to provide the knowledge that the head office needed. ING Barings, therefore, had great difficulty gathering knowledge from the different countries.

Another reason why organisations are interested in knowledge management is the growth of awareness that organisational knowledge might be the key to organisational success. Management books and articles demonstrate a growing awareness that the intellectual capital of the corporation is usually worth much more than its tangible book value (Steward, 1997; Edvinsson and Malone, 1997). Shareholders have developed a need to gain more insight in the core competence of the organisation, which in most cases resides in the (tacit) knowledge shared among the workers within the organisation. Facilitating organisational change is yet another managerial reason to engage in knowledge management. NN is a good example. Due to the internal reorganisation from product-driven to market-driven services, NN considered it necessary to capture the knowledge that already existed and make it accessible for others. At the Postbank too, it was an internal reorganisation that let to an improved structure being built for knowledge-sharing. Because managers cannot force people to share their knowledge, knowledge management calls for a fast support of knowledge workers.

Knowledge management heavily depends on the willingness of knowledge workers to take part in it. We encountered various reasons for knowledge workers to actually engage in knowledge management initiatives, such as an increase of job-efficiency, status, and fun. If the condition of a win-win situation is not taken care of, managers will be confronted with major rejections from the side of the knowledge workers.

These win-win situations do not have to match and might occur when the various actors engage in knowledge sharing out of different reasons. For example, knowledge exchanges between various actors at an electronic discussion platform introduced by the Ministry of Housing proliferated, although actors had different and sometimes even conflicting reasons to engage in these discussions.

Next to the bias towards managerial control, the management trap also relates to the bias to introduce knowledge management based on opportunity driven arguments only rather than on (present or future) problem-driven arguments. We saw that knowledge management will be more successful when it addresses existing situations and problems than when it is seen as an opportunity to organisational change.

There are basically three motives why organisations engage in explicit knowledge management activities of which the first two are expressions of opportunity driven reasons to introduce knowledge management.

One reason is ICT-driven: knowledge management is often linked to supporting knowledge exchange through ICT. With the rise of the technological possibilities that ICT offers, and especially with the rise of the intranet, Lotus Notes, and knowledge- and expert systems, new avenues opened for organisations that want to structure their knowledge processes.

Some organisations introduce knowledge management because they are triggered by stories of other organisations that engage in forms of knowledge management. As mentioned in the introduction to this paper, a possible fallacy is that most of these stories are based on conceptual orientations only or are told by highly enthusiastic (knowledge) managers. In both cases, the positive stories tend to hide negative experiences and/or pitfalls to knowledge management. In other words, organisations are seduced to imitate others, while the models they imitate are mostly incomplete. Companies that were only in a conceptual stage of introducing knowledge management, often referred to well-known textbooks, well-known best practices and well-known conference-speakers.

A third reason to introduce knowledge management is problem-driven. In such a case, organisations use knowledge management techniques to address existing or future problems. Knowledge workers themselves often initiate problem-driven knowledge management. Interestingly, all initiatives in the research that were introduced as to cope with existing problems, did not use the words "Knowledge Management" or only attached the words in a later stage when the concept gain popularity. Organisations that introduced knowledge management based on opportunity-driven arguments, all explicitly used the label but either had problems in institutionalising it or were still in a conceptual stage of introduction.

Where is knowledge managed and who's knowledge is managed? The local learning trap

Another potential problem we encountered has to do with the limited scope of attention to both process and outcomes of knowledge sharing. This so-called 'local learning trap' is a combination of two related biases: the individual knowledge bias and the operational level bias. In short, the local learning trap is about the risk to concentrate the attention on local knowledge sharing without addressing the issue how the organisation as a whole can benefit from it.

When we look at the actual practices of knowledge management and ask ourselves where knowledge is managed, we observed that many of the organisations we studied focussed their attention on the operational level only. There are various reasons why this focus on the operational level might become a burden to the knowledge management initiative. For many knowledge workers it is important that management act as an example instead of a facilitator only. As a consultant argued: "if they do not share their knowledge why would I do it?" Knowledge sharing processes cannot be limited to the operational level only. Much of the knowledge is also shared among managers. Of course, another important condition for successful introduction of knowledge management is that management not only contributes to knowledge sharing and construction, but also supports the initiatives. We did not come across this latter condition during our research, as management support was one of the criteria we used to select cases. Nevertheless, lack of management support seems to be one of the serious problems organisations face when introducing knowledge management (e.g. Davenport, 1997).

Knowledge management is generally seen as the management of learning processes within organisations. There is however a potential pitfall when this is interpreted as the management of individual learning instead of collective learning. During our research, we saw many initiatives approaching knowledge management as supporting knowledge sharing by individuals more than by collectives within organisations. That the focus tends to be more on individual learning rather than on collective learning is understandable, as managing individual learning is less complicated than collective learning. Tools to improve the individual knowledge base are part of every organisation, such as training, education, or more explicit tools such as libraries or databases. In contrast, tools to improve the collective knowledge base are much more difficult to imagine. Also, managing individual learning is easier to control than is the case with collective learning. Managers for example might ask employees to read an article, to take a course or to inspect a database. From this information processing activity, we can for a large part predict what the outcome of this learning process will be. Much of the collective knowledge is however gained during day to day interactions and is less easy to manage (Brown and Duguid, 1991). Schiphol for example had created a knowledge centre that actually functions as knowledge libraries. Individuals can acquire the necessary knowledge from these centres to gain more insights on a particular subject. The other organisations that focused its knowledge sharing on knowledge exchange: Cap Gemini, IBM and ING Barings, used the intranet to store past experiences of knowledge workers so that others could learn from this. These networks merely function as tools to support individual knowledge development more than collaborative knowledge development. As one consultant remarked: "the system is supposed to store experiences in a database, but that doesn't work, you cannot learn experiences from others as such, knowledge sharing happens through face to face communication".

Some organisations show possibilities how to avoid this individual knowledge bias. At Unilever and Stork for example the sharing of knowledge among collectives was supported by enabling the existence of communities. During frequent meetings, these communities exchange valuable experiences and develop new ideas how to improve their day to day activities. Managing communities calls for a different approach as managing individuals. In fact, management has little influence on these 'communities of practices' besides acknowledging their existence. Learning of and within communities is also often unnoticed by the learners themselves (Ciborra and Lanzara, 1994) and is seldom planned. Many communities are continuously in a flux, changing from place, time, membership and content. Mapping the knowledge within the organisation by mapping the various communities is therefore impossible; even if management is able to map all the existing communities, this would only be a random indication (Brown and Duguid, 1991). Consequently managing collective learning processes such as those that take place in communities of practices are much harder to manage than individual learning processes (Orr, 1990; Ciborra and Lanzara, 1994; Cook and Yanow, 1993; Weick and Roberts, 1993; Jordan, 1989). Because of its fluid, tacit, loose and emergent character, managing knowledge sharing by managing communities requires a different approach to management than what we are used to. This implies that the role of managers will be pushed to the periphery in which their main contribution lies in the acknowledgement and facilitation of emergent grass-root community behaviour (Wenger and Snyder, 2000; Brown and Duguid, 2000).

The most crucial consequence of the lack of management involvement is that shared knowledge will most likely remain local knowledge and will not be collectively accepted. This is not always problematic, certainly not when the knowledge is only relevant to this local group of people. In all other cases where local knowledge might be relevant to a wider context including future workers, active involvement of management to support collective acceptance will stimulate organisational learning processes. Our research showed that most initiatives focus on learning of individuals and sometimes also of groups. Seldom however is there a relation between this learning and learning at the level of organisations. We see this as a potential pitfall as most initiatives have the potential to contribute to organisational improvement.

Collective acceptance is key to bridging the gap between individual learning and organisational learning. Many of the analysed practices discussed in this paper lack this crucial process of collectively accepting shared knowledge. One of the most important reasons is the lack of collective involvement in local knowledge sharing or learning processes.

We saw that almost all companies were paying attention to supporting individual members in retrieving organisational knowledge. For some companies, such as the Postbank and the railways, this aspect of knowledge sharing was indeed their prime focus.

Many of the initiatives also involve connecting people such that individual knowledge can be exchanged. As mentioned, most companies do so by introducing electronic networks such as intranets, some rely more on physical networks such as communities and special interest groups. As such, these initiatives support either individual learning or group learning. There are however only two initiatives (IBM and NN) that support all three learning processes:

- Individual learning by giving individuals access to collective knowledge (the arrow pointing from organisational knowledge to individual knowledge);
- Individual and group learning by giving individuals and groups of individuals access to individual knowledge (the mutual arrow between individual and shared knowledge);
- Organisational learning by providing the collective access to shared knowledge (the arrow pointing from shared knowledge to organisational knowledge).

Collective acceptance at IBM mainly occurs through the intervention of a jury of specialists. This jury decides whether individually introduced knowledge in the shared knowledge database, is useable, relevant and interesting enough for the collective to be accepted for publication in the collective knowledge base. Insurance company 'National Netherlands', facilitates collective acceptance through experts, knowledge coordinators, specialists and contact persons. Because of their expertise and seniority, these people have been given the role of knowledge broker. In a way, their role is comparable with that of the jury members at IBM. Knowledge gained from contacting these people is considered as valuable and as such gets more easily accepted by the collective. The knowledge brokers are not only active in the physical networks but also play an important role in storing shared and distributed knowledge in the organisational electronic networks.

TOWARDS A SECOND WAVE OF KNOWLEDGE MANAGEMENT

We would like to refer the above discussed knowledge management initiatives as belonging to the first wave of knowledge management. One important reason why the first wave of knowledge management initiatives increasingly met with resistance is that knowledge-sharing cannot be forced; people will only share knowledge if there is a personal reason to do so. As knowledge owners, people have the power to decide if, when, how, and with whom they will share knowledge. It is an illusion to think that these decisions can be forced upon individuals. It is only when organisations acknowledge this, that the next step in the evolution of knowledge management can be made. In the second wave of knowledge management it will be the practitioners themselves who manage their own knowledge as they are in the best position to do so (see table 13.6).

Table 13.6: Differences between the first and second generation of knowledge management

Research question	1st wave	2nd wave
Why is knowledge shared?	Managerial needs	Part of daily work: as a routine
When is knowledge shared?	When there is an opportunity to do so	When there is a need to do so
Where is knowledge shared?	Operational level	Organisation-wide
Whose knowledge is managed?	Individual: human capital	Collective: social capital
What knowledge is shared?	Codified	Tacit and Codified
How is knowledge shared?	Repository systems and electronic networks.	Via personal and electronic networks

Taking seriously the power of individuals in deciding to contribute to knowledge-sharing means that organisational conditions must be changed in such a way that people would like to share. These changing conditions do not always require different organisational structures, rewards systems, positions, etc. In fact, we observed that people often do feel the need to learn and share knowledge with others in situations where this would help them to do their work better, more efficiently and with more satisfaction. This certainly applies to situations in which knowledge-sharing contributes to the daily operations of the organisation. All cases dealing with ICT tools to support knowledge retrieval were successful in the sense that they formed an integral part of the daily practices. That they were successful has much to do with the fact that the tools contributed to people's work. Although in some cases extra effort was required, this was outweighed by the benefits of work improvement.

People do want to share knowledge but only if there are good reasons to do so. Personal triggers to share knowledge are, for example, when it provides recognition from significant others, when it contributes to daily practices, or when it contributes to individual learning processes. The most important obstacle to managing knowledge is management itself. The role of managers in the next generation of knowledge management will be much more on the periphery, providing opportunities for people to exchange knowledge. What first needs to be addressed is the question how to stimulate a need to share knowledge among a group of people. It is only when this need is satisfied, that physical or electronic spaces are used for knowledge-sharing purposes. The second stage in the development of knowledge management places this need for knowledge connections central. For this purpose, authors within the area of organisation and management increasingly start to link the idea of social capital with knowledge management (e.g. Cohen and Prusak, 2000; Lesser, 2000; Nahapiet and Ghosal, 1998).

The notion of 'social capital' should be seen as an additional ingredient to the already well-known economic conditions or elements that make up organisational capital: physical capital, financial capital, and human capital. Where human capital refers to individual ability (Becker, 1964), social capital refers to social networks that create opportunities. While the notion of human capital formed the core knowledge of the first wave of knowledge management, social capital can be seen as the core ingredient of the second wave of knowledge management. Together, human and social capital make up the intellectual capital of the organisation. Human capital relates to individual learning but does not necessarily contribute to organisational learning. It is argued that social capital provides the conditions that nurture a willingness among these intellectual humans to connect (Adler and Kwon, 2002).

Investing in social capital means that long term benefits such as social networks based on reciprocity, trust, and mutual respect and appreciation will last much longer than engineered networks such as organisational teams (Nahapiet and Goshal, 1998). Nahapiet and Goshal (1998) argue that social capital has three dimensions that are highly interrelated and difficult to segregate in practice:

- a **structural dimension** such as network ties, network configurations and appropriable organisation (e.g. Burt, 1992; Granovetter, 1992; Coleman, 1988),
- a **cognitive dimension** such as shared codes and language, and shared narratives (e.g. Cicourel, 1973; Orr, 1990), and

- a **relational dimension** such as mutual trust, norms, obligations and identification (e.g. Fukuyama, 1995; Cohen and Prusak, 2001).

Investing in social capital seems even more important with the growth of virtual organisations, e-lancers and geographically dispersed settings together with and the continuous changes that are a result of shifting partnerships and boundaries. Due to these organisational changes, existing social capital can easily be shaken up.

One way to help people connect despite geographic and time differences, is through the use of communication technology, such as e-mail, telephone, video conferencing and more advanced groupware technologies. It has already been argued for many years that GroupWare will be a very important tool in present and future organisational settings. However, so far, empirical studies that suggest the institutionalisation of these tools in the group processes have been remarkably limited (e.g. Bowers, 1994; Orlikowski, 1996; Ciborra, 1996; Zuboff, 1988). There could be many explanations for this, but given what our cases tell us, perhaps the most convincing one is that people do not use these sophisticated tools simply because their need to share knowledge is limited. It might well be that the use of GroupWare tools will start to increase the moment the value lies more in the network than in individuals.

The role of ICT in social capital can be seen as bi-directional (Huysman and Wulf, in press). A high level of social capital, shown for example by pre-existing strong non-electronic networks, is a success factor in establishing electronic based networks (Fukuyama, 1995). This is also what we saw happening in the cases discussed in this chapter. But at the same time, the existence of ICT possibilities might create a networking infrastructure, which encourages the formation of social capital (Calabrese and Borchert, 1996). It is thus an empirical question which tendency will dominate.

To conclude, we have argued that sharing knowledge is a collective rather than an individual activity, that it will only occur naturally in situations where individuals benefit from sharing and that ICT can only support it, not replace it. We showed how managing knowledge sharing cannot rely solely on knowledge management tools, nor on reward structures, formal knowledge management strategies, chief knowledge officers, training programmes etc. Our critical observations about knowledge management might have given the reader a rather negative impression. Nevertheless, this should not imply that we do not see a future for managing knowledge sharing. On the contrary, there are various reasons why managing knowledge sharing is extremely important. In fact, we are so convinced of the importance of managing knowledge sharing that we believe organisations should strive to make it a routine part of their daily activities.

REFERENCES

Adler and Kwon *(2002)*. Social capital, *Academy of Mangement Review*

Becker, G. (1964). *Human capital, a theoretical and empirical analysis with special reference to education.* Columbia University, New York

Bowers, J. (1994). The work to make a network work: studying CSCW in action. In: *Proceedings of CSCW'94*, Chapel Hill, NC: ACM Press.

Brown, J. S. and P. Duguid (1991). Organizational learning and communities-of-practice. Towards a unified view of working, learning, and innovation. *Organization Science, 2*, 40-57.

Brown, J. S. and P. Duguid (2000). *The social life of information*, Harvard Business School Press, Cambridge MA.

Burt, R. S. (1992). *Structural holes: the social structure of competition*, Harvard University Press, Cambridge.

Calabrese, A. and M. Borchert (1996). Prospects for electronic democracy in the United States: Rethinking communications and social policy. *Media, Culture, and Society, 18*, 249-268.

Ciborra, C. (eds) (1996). *Groupware and teamwork. Invisible aid or technical hindrance?* Wiley, New York.

Ciborra, C. U. and G. F. Lanzara (1994). Formative contexts and information technology, understanding the dynamics of innovation in organizations. *Accounting, Management and Information Technology, 4* (2).

Cicourel, A. V. (1973). *Cognitive sociology: Language and meaning in social interaction.* Penguin Press, New York.

Coleman, J. S. (1988). Social capital in the creation of human capital. *American Journal of Sociology* (Supplement) **94**, pp. 95-120.

Cook, S. D. N. and D. Yanow (1993). Culture and organizational learning. *Journal of Management Inquiry, 2* (4).

Davenport, T. H. (1997). Ten principles of knowledge management and four case studies. *Knowledge and Process Management, 4* (3).

Edvinsson, L. and M. S. Malone (1997). *Intellectual Capital: realizing your company's true value by finding its hidden roots.* Harper Business, New York.

Fukuyama, F. (1995). *Trust: the social virtues and the creation of prosperity.* Free Press, New York.

Granovetter, M. (1985). Economic action and social structure: The problem of embeddedness. *American Journal of Sociology,* **91**, 481-510.

Huysman, M. H. and D. de Wit (2002). *Knowledge sharing in practice.* Kluwer Academic Publishers.

Huysman, M. H. and V. Wulf (in press). *Social capital and information technology.* MIT Press: Cambridge

Jordan, B. (1989). Cosmopolitical obstetrics: some insights from the training of traditional midwives. *Social Science and Medicine,* **28** (9).

Lave, J. and E. Wenger (1991). *Situated learning: Legitimate peripheral participation.* Cambridge University Press: Cambridge.

Lesser, E. L. (ed) (2000). *Knowledge and social capital: Foundations and applications.* Butterworth Heinemann., Boston.

Nahapiet, J. and S. Ghoshal (1998). Social capital, intellectual capital, and the organizational advantage. *Academy of Management Review,* **23** (2), 242-266.

Orlikowski, W. (1996). Evolving with notes, organizational change around groupware technology. In: *Groupware and teamwork* (C. U Ciborra, ed.). Wiley, New York.

Orr, J. E. (1990). Sharing knowledge, celebrating identity: Community memory in a service culture. In: *Collective remembering: memory in society,* (P. Middleton and D. Edwards, eds.). Sage, London.

Steward, T. A. (1997). *Intellectual capital: the new wealth of organizations.* Doubleday, New York.

Weggeman, M. (1997). *Kennismanagement: Inrichting en besturing van kennisintensieve organisaties.* Scriptum, Schiedam, The Netherlands

Weick, K. E. and K. H. Roberts (1993). Collective mind in organizations: heedful inerrelating on flight desks, *Administrative Science Quarterly,* **38** (3), 357-381.

Wenger, E. and W. M, Snyder (2000). Communities of Practice: The Organizational Frontier. *Harvard Business Review,* **78**, 139-146

Wiig, K. M. (1997). Integrating intellectual capital and knowledge management. *Long Range Planning,* **30** (3), 399-405.

Zuboff, S. (1988). *In the age of the smart machine.* Basic books, New York

14

Knowledge Management – Consequences from a Study on Knowledge Management Systems in Top 500 German Companies

Roland Maier and Franz Lehner[1]

Goals of the Study and Research Design

Little is known about how knowledge management systems (KMS) are applied in organisations and what results can be seen in terms of organisational effectiveness. There have been a number of studies, mainly in the US, on the application of knowledge management in organisations. So far, the studies in general either distilled "best practices" out of a number of "success stories" (case studies) or studied the notion of knowledge management in a very broad and general way (see e.g. Bullinger *et al.,* 1997; Davenport *et al.,* 1998; ILOI, 1997, for an overview see Lehner, 2000; Maier 2002). However, none of them focussed on the technological support for knowledge management – KMS – without neglecting the other important points of intervention of a knowledge management effort, namely people, organisational design and culture.

In 1999, an extensive study of the state-of-the-art investigating the use of knowledge management systems in big German companies and the development of concepts, scenarios and reference models for the management of KMS in organisations has been carried out at the Institute for Business Informatics of the University of Regensburg. Figure 14.1 outlines the underlying research model. In the following, selected parts of this model are discussed in

---

[1] University of Regensburg, Germany

detail. For a detailed description of the dimensions and variables and the theoretical framework see Maier and Klosa (1999) and Maier and Lehner (2000), Maier (2002).

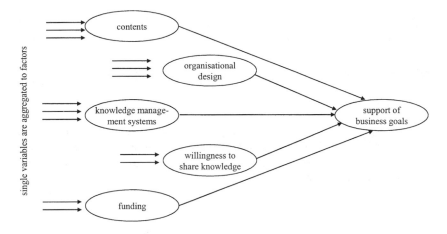

Figure 14.1: Research model

The research model structures the organisational memory into organisational memory contents and the concept of the application of KMS. The OM contents consist of the knowledge of the members of the organisation, i.e. knowledge in peoples' minds or documented knowledge which can be paper based and/or in electronic form. The contents can be structured according to an organisation-wide knowledge structure (e.g. tree, semantic net). The the concept of the application of KMS in turn influences how the organisation deals with its content. The concept of the application of KMS consists of the organisational design of KMS use (structuring of tasks and processes, roles, scope of the application of KMS), organisational culture (especially values, rules and norms concerning the willingness to share knowledge) and knowledge management systems. The concept is the main unit of analysis in this study. This concept is managed by a knowledge management unit which sets the goals for the concept of the application of KMS. It is dependent on the business environment, especially on the general organisational design (e.g. degree of centralisation), the size and sector of the organisation. The concept is also dependent on the funding provided and will result in a certain degree of organisational effectiveness as well as results concerning management and handling of knowledge in the company (which is captured by a number of OM measures). The application of the concept should support and positively impact business goals. The arrows in figure 14.1

show on the one hand the dependencies between these concepts as hypothesized in the study and on the other hand that each of the concepts is measured by a number of individual variables (see Maier 2002 for a detailed description of these variables).

There are a number of goals that companies can direct their KM efforts to. Generally there are two different strategies which can be applied in the implementation of knowledge management in companies: the codification and the personalisation strategy (see Hansen *et al.,* 1999, 109). The codification strategy focuses on the documentation and institutionalisation of (explicit) knowledge (see e.g. Zack, 1999) who defines a framework for the management of explicit knowledge and expertise). The personalisation strategy supports the direct communication link between individual (human) experts and knowledge users. The list of goals the importance and achievement of which is assessed subjectively in our empirical study covers both of these strategies. The list is derived from case studies documented in the literature (see e.g. Davenport *et al.,* 1998, who derive a list of objectives of knowledge management projects) as well as from empirical data found in studies focussing broadly on knowledge management (e.g. ILOI, 1997; Bullinger *et al.,* 1997):

- identify existing knowledge/make existing knowledge transparent,
- improve documentation of existing knowledge (both in terms of the quality of the content and structure),
- change (parts of) the organisational culture (e.g. willingness to share knowledge),
- improve communication and co-operation (both, within and between formal work groups/teams),
- turn implicit, "subjective" knowledge into explicit, "objective" knowledge,
- improve education, training and networking of newly recruited employees (both, job starters, such as trainees, apprentices, graduates, and newly hired experienced employees, experts),
- improve training and education of all employees (personnel development),
- improve retention of knowledge (e.g. in the case of employees leaving the organisation),
- improve access to existing sources of knowledge,
- improve acquisition or purchasing of external knowledge,
- improve distribution of knowledge,
- improve management of innovations (= research and development, e.g. more innovations, patents, faster innovations, avoid multiple development of the same concept),
- reduce costs (e.g. reduce organisational redundancy, reduce the use of paper, reduce travel expenses),
- sell knowledge (e.g. licensing, consulting, access to KMS).

Knowledge management is supposedly an ongoing effort in organisations which is not a completely new phenomenon. All the tasks related to knowledge management and carried out in an organisation are called the knowledge management function. Successful organisations have always organised their knowledge resources efficiently (see Roehl, 1999, 13). However, in many organisations the relevant activities have rested in the hands and minds of talented individuals. New is the systematic approach to the management of the knowledge resource which requires organisational design. Therefore, we examine who (which role) is responsible for what knowledge management tasks. The list of knowledge management tasks does not cover all tasks thinkable (see e.g. Probst *et al.,* 1997; Schüppel, 1996). The functions are derived from the definition of knowledge management systems and a model of organisational information processing:

- knowledge identification,
- acquisition of external knowledge,
- release of knowledge elements (formal approval of institutionalisation),
- storing of knowledge elements,
- knowledge classification,
- updating/extending of existing knowledge structure (ontology),
- knowledge distribution,
- knowledge quality management, refinement, repackaging of knowledge, knowledge deletion, archiving, knowledge selling.

There are a great number of systems on the market which claim support for organisational memory, organisational learning or knowledge management respectively. The field is still immature, though, in the sense that there are no classes of systems that the literature has agreed on. So far there are several proposals for classifications of systems which mostly lack completeness and also exclusiveness in the sense that one system fits into one and only one category (see e.g. Malloy, 1998, 3-5; Jacobsen, 1996, 169; Stein and Zwass, 1995, 91; Ackerman, 1994, 10ff). Thus, it is not surprising that the systems on the market are more or less sets of functions thought to be useful for knowledge management.

Therefore we decided to focus on the functions provided by systems as different as Intranet platforms, group support systems, communication systems or systems explicitly described as knowledge management systems. A list of 66 functions was derived a) from an extensive survey of existing knowledge management systems, b) from a set of empirical studies on knowledge management and c) from approaches to classify this kind of systems in the literature.

The following groups of functions were identified (for a comprehensive list of the functions see Maier and Klosa, 1999):

- support of knowledge search and presentation
- support of knowledge publication, structuring and feedback
- support of communication and co-operation
- support of computer based teaching and learning
- support of the administration of knowledge management systems

Generally, both, normative suggestions for KMS and actual implementations of KMS vary considerably in terms of the content to be managed. Many companies seem to be driven by a pragmatic approach which puts those parts of the organisational knowledge at the centre of consideration the management of which would promise the most direct positive effects. Examples are patents, skills data bases (yellow pages), lessons learned, best practices, descriptions of products, processes or the structural organisation and the like. In many cases, explicit knowledge is predominant.

Knowledge management efforts vary not only in terms of goals and organisational design, but also in terms of size and funding. Three concepts of funding KM activities can be differentiated. (1) Most organisations supposedly finance knowledge management efforts in terms of a budget allocated to a group or a project. This is due to various reasons. First of all, knowledge management is a rather new approach which is propagated to increase its acceptance throughout the organisation. In this first phase, the knowledge management efforts are funded centrally and the usage of the corresponding services is open to all departments, processes or individuals. (2) As the concept matures, most organisations will try to at least allocate costs where they are generated which means that services are charged for. (3) The final step might be a move to a market scheme where demand and supply of knowledge management products and services are brought together, both internally within the organisation and externally with business partners.

One of the most prevalent questions in the knowledge management area widely discussed in literature and practice is how do we determine the value created and the benefits gained by the application of such efforts (see e.g. Stuart, 1996, 2). Considering the fact that there is still considerable disagreement about what exactly knowledge is or knowledge resources are which have to be managed (for an extensive critic see Roehl, 1999) it is difficult to assess what the results of the application of such a concept would be and especially what the differences to not applying this concept would be. Several approaches to this problem can be distinguished, e.g. the Intellectual Capital (IC) approach (see Wiig, 1997; Stewart, 1997).

KMS IN GERMAN ORGANISATIONS – SELECTED RESULTS OF THE STUDY

In the empirical study the following methods were used: exploratory unstructured interviews, questionnaire, telephone interviews and personal interviews.. As target organisations a list of the top 500 German companies and the top 50 banks and insurancies was used. The overall sample size was 504. The number of respondents was 73 (response rate 14.5 %). In the following, a brief outline of selected results of the study concerning KMS is given. A more extensive discussion as well as the full details of the study can be found in Maier (2002).

Most large organisations have an Intranet and/or a Groupware platform in place that offers basic KM functionality and a solid foundation for KMS

By now, almost all large organisations have installed an Intranet and/or a Groupware solution which can be considered the basic information and communication technology (ICT) infrastructure for KM. These platforms together with a multitude of extensions and add-on tools provide good, basic KM functionality. During the past couple of years, corporate Intranet solutions have been implemented to connect employees, to support the easy sharing of electronic documents and to support access to company information. Also, organisations have installed Groupware tools in order to support teams and to master the increasing complexity of organisational structure and processes along with advanced information and communication needs.

Many KMS functions are implemented, but not used intensively

Large organisations have already implemented many KM-specific functions as part of advanced Intranet infrastructures and Groupware platforms as well as more specific solutions, such as customer relationship management systems or systems that support individual business units. Many of the functions are not used intensively, in some cases due to technical problems, but mostly because they require substantial organisational changes. Therefore, there still seem to be considerable potentials when applying information and communication technology to KM initiatives.

Integrative KMS functions predominate, but interactive and bridging KMS functions catch up

The distinction between integrative and interactive functions was made by Zack (1999) and corresponds to the two knowledge management strategies codification and personalization (Hansen *et al.,* 1999). Integrative functions aim at combining and integrating heterogeneous sources of documented knowledge. Interactive functions support the location of and communication, coordination and cooperation between knowledge providers and knowledge seekers. Up to now, in most organisations there has been a strong emphasis on integrative KMS functions with a focus on explicit, documented knowledge. This is not surprising as in many cases large amounts of documents have already existed in electronic form. The improved handling of documents and the redesign of business processes to systematically capture lessons learned and to use the document base have provided for a visible improvement of the organisation's knowledge base. Recently, there is a trend towards more collaboration-oriented and bridging KMS functions. Bridging functions aim at bridging the gap between integrative KMS functions and their concentration on codified, documented knowledge with interactive KMS functions that allow for direct sharing of knowledge from person-to-person. Organisations profit from integrative KMS functions and now seek for new forms of information and communication technology support for their KM initiatives. Also, the technical requirements for a sophisticated support of media-rich electronic communication and collaboration can now be met at a reasonable cost due to the advancements in the information and communication technology infrastructure in the organisations. Examples are videoconferencing, teleteaching and telelearning or application sharing that require large bandwidths and multimedia equipment for the PCs of the participating knowledge workers. Most organisations follow a general pattern of four phases in which they implement predominantly (1) basic KM-related functionality, (2) integrative KMS functions, (3) interactive KMS functions before they (4) finally aim at a combination and integration of the three. Most organisations are still in the first two phases of this sequence whereas many organisational KM instruments need to be complemented by KMS functions of the third and fourth phase.

KM-related information and communication technology systems lack integration

In most organisations, a multitude of partial systems are developed without a common framework which could integrate them. Only recently, comprehensive and integrated KMS gain market share. They offer extensive functionality integrated within one system. Some organisations also build enterprise knowledge portals that at least integrate access to most, if not all organisational and organisation-external information and communication technology systems relevant for the KM initiative. Still, in most organisations the functionality of KM-related ICT systems is largely not integrated, e.g., messaging systems, document or content management systems, access to external systems, World Wide Web, external online data bases, data warehouses, customer relationship management systems and last but not least the organisation's enterprise resource planning systems.

KMS are highly complex systems

Comprehensive KMS are highly complex ICT systems because of (1) the technical complexity of the "intelligent" functions that distinguish a KMS from a more traditional system and of the large volumes of data, documents and messages as well as links, contextualisation and personalisation data that have to be handled, (2) the organisational complexity of a solution that affects business and knowledge processes as well as roles and responsibilities throughout the organisation and finally (3) the human complexity due to the substantial change in the handling of knowledge that is required from the organisation's knowledge workers as KMS have to be integrated into their work environment.

Most organisations build their own KMS solutions

The majority of organisations relies on organisation-specific developments and combinations of tools and systems rather than on standard KMS solutions available on the market. The most important explanations for this finding might be two-fold. On the one hand, the market for KMS solutions is a confusing and dynamic one. There is no leading vendor or group of vendors yet and interoperability with other KM-related systems that the organisations have in place is often difficult to realise. On the other hand, organisations might fear that they loose strategic advantages if they exchange their home-grown organisation-specific KMS solutions for standard software that might not fit their needs as well. However, more recently, the market for standard KMS solutions has consolidated so that more and more organisations rely on standard solutions (e.g., OpenText Livelink, Hyperwave e-Knowledge Server, Verity Knowledge Suite) that they customize to their needs.

The diversity of KMS contents has increased

Generally, more organisations handle a larger variety of knowledge contents when compared to previous studies. About half of the organisations use modern KM contents, like employee yellow pages, skills directories, idea and proposal systems and lessons learned. Recently, organisations seem to have extended the scope of their KMS to include more types of internal knowledge previously unavailable to a larger group of employees. Most organisations have organisation-specific descriptive knowledge on the one hand and unapproved contributions to knowledge networks on the other hand as part of their knowledge base. Secured inventions (e.g., patents) are used by only a minority of organisations. The biggest potentials seem to lie on the one hand in experiences and expertise that bridge the gap between organisation-specific descriptive knowledge and unapproved contributions in knowledge networks. On the other hand, external knowledge bridges the gap between the organisation and its environment. Many organisations do not distinguish between these KM-related contents and more traditional contents of ICT systems, such as a broad view of all documents or the entire content of the corporate Intranet, data in data warehouses or transactional and communication data about customers and business partners. There is still considerable uncertainty in many organisations about what is or what should be considered a knowledge element in an organisation's KMS.

CONCLUSIONS

By now, almost every big organisation like the ones in the sample of the study has installed an Intranet and/or a Groupware solution, which can be considered the basic information and communication infrastructure for KM. Many organisations apply multiple platforms, e.g., Lotus Notes and Intranet server. Due to the integration capabilities of the Web technologies, it is technically feasible to integrate the platforms, use a combination of those functions of each platform that are needed and to hide this added complexity from the participants.

These platforms together with a multitude of extensions and add-on tools provide good KM functionality. Almost two thirds of the organisations have developed their own KMS solutions based regularly on a bundle of tools. It is not surprising that in most of the organisations KM functionality is spread over many systems with a varying degree of integration. However, about a third of the organisations reported a high degree of integration with all functions at least integrated within one single user interface or fully integrated within one system.

According to the results of the empirical study and especially the interviews with organisations pioneering the implementation of KMS, there seem to exist general patterns describing the sequence or phases of implementation or evolution of KMS functions in organisations (cf Maier 2002):

1.   basic KM-related functionality:
Groupware platforms and Intranet solutions provide basic interactive functionality (email, email distribution lists) and integrative functionality (publishing, search and retrieval of documents) which is used intensively.

2.   integrative KMS:
Advanced KM functionality is implemented to support the codification of knowledge and search and retrieval as well as the administration of knowledge repositories and the organisation of knowledge structures.

3.   interactive KMS:
Sophisticated KM functionality supports the location of experts, their communication and collaboration, provides shared workspaces for communities, and modern e-learning instruments.

4.   bridging KMS:
Finally, integrative and interactive KMS are combined to provide highly contextualised knowledge repositories which also focus on linking knowledge seekers and providers, match participants with similar profiles, make recommendations, filter and present knowledge elements and links in a personalised way.

Not surprisingly, most organisations surveyed are still in the first two phases of KMS implementations. Many critics of using information and communication technology to support KM also only consider functions of these two phases and might not be aware of the potentials of the functions of the higher phases. It is precisely the more advanced integrative, the interactive and particularly the bridging KMS functionality that overcome the problems of the document-oriented, technical KMS solutions that do not fit organisational instruments.

Most organisations seem not to keep track of the use of KMS and KM-related services. Even in those cases in which usage figures were evaluated, it would be difficult to judge whether the users of KMS actually found and could apply the knowledge that they looked for. Consistently with other KM studies, improved speed of innovation is an important business goal supported by KM. In addition to this rather KM-specific goal, organisations seem to primarily target the same business goals as used in BPR or process management projects: improve customer satisfaction, improve productivity and improve scheduling. Improve growth of organisation was ranked lowly in all KM studies reflecting once again the internal focus of most KM initiatives. KM initiatives attempt to improve primarily the organisations internal way of handling knowledge in order to achieve traditional business goals, oriented towards value creation, rather than environment-oriented goals such as improve growth, reduce risks and develop new business fields.

The following theses present some final recommendations, which reflect expected potentials and pitfalls concerning KMS:

Critical factors for the implementation of KMS are the following (see also Kortzfleisch and Winand, 1997):

- ease of use (provided by a simple user interface),
- user integration and user participation,
- integration of existing (legacy) applications,
- technical security and reliability,
- support functionality (including „meta-knowledge") so that the user is able to use the KMS autonomously (please explain),
- applications that support more (inter-) active use of the medium
- the suitability of the organisation or management concept respectively (see also thesis 3).

Some authors postulate a change to a more trustful organisation culture, in which "information democracy" stimulates the willingness of employees to learn and to provide their knowledge for others. In this context, KM solutions with their tendency to a greater degree of information sharing can be successful. Thus, it is not surprising that security issues play an important role, so that the employees trust the KM system.

Like the unmanaged Internet one could suppose that an unmanaged Intranet would grow into a dynamic and chaotic massive information store so that „Intranet mining" methods and tools have to be developed to support the tasks of information resource discovery, information extraction and use. As a consequence, some kind of management (e.g. information or knowledge desk) is needed to prevent „information overload". However, note that a central „authority" would counteract the open and less hierarchical culture of an Intranet.

The design perspective in a KMS development has to be a more holistic one, which means focused on organisational development as opposed to conventional software development where the design perspective is normally task- or process-oriented. Thus, the success or benefit related to the implementation of KMS cannot be measured by comparing task performance before and after the installation of the system. Therefore, the funding of KMS development projects has to be on the enterprise level, not on the departmental (or any other lower organisational) level.

New professions have to be created, such as change agent, knowledge architect, KM administrator and the like, who should be responsible for the dissemination of KM ideas throughout the whole organisation.

The information processing view, evident in many definitions of knowledge management in the press and academic texts, has often considered organisational memory of the past as a reliable predictor of the dynamically and discontinuously changing business environment. Most such interpretations have also made simplistic assumptions about storing past knowledge of individuals in the form of routinised programmable logic, rules-of-thumb and archived best practices in data bases for guiding future action. However, there are major problems that are attributable to the information-processing view of information systems. These problems are finally described as three key myths about knowledge management which can be found in several variations in literature.

Myth 1: Knowledge management technologies can deliver the right information to the right person at the right time. This idea applies to an outdated business model. Information systems in the old industrial model mirror the notion that businesses will change incrementally in an inherently stable market, and executives can foresee change by examining the past. The new business model of the Information Age, however, is marked by fundamental, not incremental, change. Businesses can no longer plan long-term; instead, they must shift to a more flexible "anticipation-of-surprise" model. Thus, it is impossible to build a system that predicts who the right person at the right time even is, let alone what constitutes the right information.

Myth 2: Knowledge management technologies can store human intelligence and experience. Technologies such as databases and groupware applications store bits and pixels of data, but they cannot store the rich schemas that people possess for making sense of data bits. Moreover, information is context-sensitive. The same assemblage of data can evoke different responses from different people. Even the same assemblage of data when reviewed by the same person at a different time or in a different context could evoke differing response in terms of decision-making and action. Hence, storing a static representation of the explicit representation of a person's knowledge ~ assuming the willingness and the ability to part with it - is not tantamount to storing human intelligence and experience.

Myth 3: Knowledge management technologies can distribute human intelligence. Again, this assumes that companies can predict the right information to distribute and the right people to distribute it to. And bypassing the distribution issue by compiling a central repository of data for people to access does not solve the problem either. The fact of information archived in a database does not ensure that people will necessarily see or use the information. Most of our knowledge management technology concentrates on efficiency and creating a consensus-oriented view. The data archived in technological "knowledge repositories" is rational, static and without context and such systems do not account for renewal of existing knowledge and creation of new knowledge.

Given the dangerous perception about knowledge management as seamlessly entwined with technology, "its true critical success factors will be lost in the pleasing hum of servers, software and pipes" as observed by the author's interviewee in a recent CIO Enterprise magazine. A few years ago, technologies such as Intranet technologies, Lotus Notes, MS-Exchange were being considered as enablers of knowledge management. The more recent interest is on technologies related to knowledge portals and still-on-horizon products. Despite significant advancement in technologies and substantial investment by companies in such technologies, most organisations are still trying to find answers to simple questions such as: How to capture, store and transfer knowledge? How to ensure that knowledge workers share knowledge? Given the quest for answers to such questions, it becomes imperative for organisations to clearly understand the above strategic distinction between knowledge and information.

This strategic difference is not a matter of semantics, but has critical implications for managing and surviving in an economy of information overabundance and information overload. As most new media and Net executives competing for 'eyeballs', 'mindshare', and virtual communities, would realise, in the new world of e-business, knowledge technology and information society, the scarce resource is not information, but human attention.

REFERENCES

Ackerman, M. S.(1994). Definitional and contextual issues in organizational and group memories. In: *Proceedings of the 27th Hawaii International Conference of System Sciences (HICSS)*, Organizational Memory Minitrack, January 1994

Bullinger, H.-J., K. Wörner and J. Prieto (1997). Wissensmanagement heute. Daten, Fakten, Trends. Fraunhofer Institut Arbeitswirtschaft und Organisation, Stuttgart 1997

Davenport, T. H., D. W. de Long and M. C. Beers (1998). Successful knowledge management projects. *Sloan Management Review*, **39** (2), 43-57.

Hansen, M. T., N. Nohria and T. Tierney (1999). What's your strategy for managing knowledge?. *Harvard Business Review,* **77** (2), 106-116.

Internationales Institut für Lernende Organisation und Innovation (ILOI) (1997). *Knowledge Management – Ein empirisch gestützter Leitfaden zum Management des Produktionsfaktors Wissen.* München.

Jacobsen, A. (1996). Unternehmensintelligenz und Führung "intelligenter" Unternehmen. *technologie & management*, **45**, (4), 164-170.

Kortzfleisch, H. von, and U. Winand (1997). Kooperieren und Lernen im Intranet. *Information Management*, **2**, 28-35

Lehner, F.(2000). *Organisational Memory. Konzepte und Systeme für das organisatorische Lernen und das Wissensmanagement.* Munich, Vienna.

Maier, R. (2002). *Knowledge management systems. Information and communication technologies for knowledge management.* Berlin.

Maier, R. and O. Klosa, (1999). Knowledge management systems '99. State-of-the-art of the use of knowledge management systems, **1** – Design of an empirical study. Research paper no. 35, University of Regensburg, Department of Business Informatics III.

Maier, R. and F. Lehner (2000). Perspectives on knowledge management systems. – Theoretical framework and design of an empirical study. In: *Proceedings of the European Conference on Information Systems – ECIS'2000* (H. R. Hansen, M. Bichler and H. Mahrer, eds.), Vol. 1, 685-693. Wirtschaftsuniversität Wien, Wien 3. – 5. Juli.

Malloy, A. (1998). Supporting knowledge management: You have it now. *Computerworld,* February 23, 1998.

Probst, G. J., S. Raub and K. Romhardt (1997). *Wissen managen. Wie Unternehmen ihre wertvollste Ressource optimal nutzen.* Wiesbaden: Gabler.

Roehl, H. (1999). Kritik des organisationalen Wissensmanagements. *Organisationslernen durch Wissensmanagement* (Projektgruppe wissenschaftliche Beratung, eds.), pp. 13-37 :Frankfurt/Main.

Schüppel, J. (1996). *Wissensmanagement: organisatorisches Lernen im Spannungsfeld von Wissens- und Lernbarrieren.* Wiesbaden 1996.

Stein, E. and V. Zwass (1995). Actualizing organizational memory with information systems. *Information Systems Research*, **6** (2), 85-117.

Stewart, T. A. (1997). *Intellectual capital: The new wealth of organisations.* New York.

Stuart, A.(1996). 5 uneasy pieces – part 2: Knowledge management. *CIO Magazine*, June 1st, 1996, URL: http://www.cio.com/archive/060196_uneasy_1.html/

Wiig, K. M. (1997). Integrating intellectual capital and knowledge management. *Long Range Planning*, **30** (3), 399-405.

Zack, M. H. (1999). Managing codified knowledge. *Sloan Management Review*, **40** (4), 45-58.

PART 4

BARRIERS AND CONDITIONS

15

Task Orientation Matters: Knowledge Management from an Individual Level Perspective

Sabine Sonnentag[1]

Introduction

In the past, both researchers and practitioners have stressed the importance of knowledge management and suggested various approaches for improving knowledge management in organisations. Within this discussion, relative little attention has been paid to the tasks to be accomplished. However, individual knowledge management can only be successful if it matches the knowledge requirements of the tasks and individuals' willingness to acquire, use, and distribute knowledge prior to and during the task accomplishment process. This chapter presents a theoretical model on individual knowledge management activities. Basically, it suggests that individuals' task orientation is a crucial factor for knowledge management.

Knowledge and Knowledge Management in Today's Organisations

Knowledge is an increasingly important resource in organisations necessary for succeeding in a continuously changing, global economy (Davenport and Prusak, 1998). Moreover, knowledge becomes obsolete very quickly and has to be updated regularly (Pazy, 1994). Therefore it

[1] Technical University of Braunschweig, Germany

becomes one important task for organisations to integrate new knowledge both from inside and outside its boundaries and to distribute it among their employees on a regular basis. To ensure the integration, distribution, and application of knowledge in organisations, both researchers and practitioners express great hopes with respect to the benefits of a systematic knowledge management approach (Grant, 1996; Nonaka, 1994; Probst, Raub, and Romhardt, 2000).

Knowledge management comprises processes and procedure that aim at an improvement of the creation, exchange, and use of knowledge (DeLong and Seemann, 2000). It includes a broad range of activities such as knowledge identification, knowledge acquisition, knowledge development, knowledge sharing and distribution, knowledge utilisation, and knowledge retention (Probst *et al.*, 2000).

Additionally to the increased importance of knowledge as a resource, organisations and their employees suffer from information overload (Sparrow and Daniels, 1999). Employees are faced with great amounts of information – a trend which is well illustrated by the increased amount of memos, email and voice-mail messages circulating within and between today's organisations. Often, employees feel overtaxed by this information and feel unable to identify and select the most relevant pieces of information. Information overload threatens individual well-being and performance as well as organisational effectiveness. One can speculate that information overload will increase in the near future, particularly in the case of lacking or inappropriate knowledge management.

To deal with the problems associated with the amount, importance and development of know-ledge, organisations chose different strategies. One of these strategies stresses codification, the other stresses a more personalized approach that centers on the exchange of knowledge between individuals through direct interpersonal interaction (Hansen, Nohria, and Tierney, 1999). Knowledge management as discussed within this chapter can be subsumed under such a personalisation strategy. Thus, the chapter addresses the activities performed by organisational members in order to acquire and share knowledge necessary for task accomplishment. More specifically, I approach personalized knowledge management from an individual level perspective. This approach implies the deliberate focus on knowledge management activities of individuals, i.e. activities performed by individuals in order to enable or improve their own task accomplishment process or their co-workers' task completion processes. These activities may become obvious as acts of interpersonal behaviour (e.g., asking a co-worker, passing on information to another person) as well as acts that are performed in individual settings (e.g., consulting a knowledge data-base). More specifically, individual knowledge management activities comprise (1) knowledge acquisition, including searching for knowledge, (2) distribution of knowledge, and (3) storage of knowledge.

My focus on individual knowledge management differs from other approaches that focus on knowledge management performed at and directed to the team or organisational level. In organisational practice, however, these individual knowledge management activities do not occur in isolation. Through externalisation and knowledge exchange, individual knowledge benefits from and contributes to shared knowledge. Organisational and external knowledge results in individual knowledge through inclusion and internalisation processes (see Huysman, chapter 3 in this book).

I assume that these individual level activities are highly relevant – although not sufficient – for knowledge management at higher organisational levels (e.g., team level, company level). Organisationally relevant processes such as organisational knowledge creation are achieved through individual level activities, e.g., individuals' knowledge creation activities (Nonaka, 1994). As other organisational features influence organisational effectiveness through individual-level processes (Tesluk, Hofmann, and Quigley, 2002), also organisational knowledge management concepts and interventions influence organisationally desired outcomes through individuals engaging in knowledge management. Therefore, it is highly important to identify the prerequisites of individual knowledge management. In this chapter, I describe some of the core factors necessary for an individual to pursue knowledge management activities. Particularly, I argue that the task to be accomplished and an individual's task orientation are crucial for knowledge management and the prevention of information overload.

TASK ORIENTATION, KNOWLEDGE MANAGEMENT BEHAVIOUR, AND INFORMATION OVERLOAD

The Task Orientation Concept

Task orientation is an ambiguous concept within the psychological literature (Fleishman, 1953; Nicholls, 1984; Ulich, 1991). I define task orientation as an individual work-related emphasis on the attainment of one's task goals. High task orientation is characterized by a focus on the very task and its related goals. In case of high task orientation, an individual prioritises task-related goals in his or her action process over other goals not primarily task-related (e.g., ego-related goals). Additionally, in case of high task orientation an individual will pursuit other work-related activities (e.g., exchanging information with co-workers, attending training programs) in the light of his or her work tasks. The guiding principle in deciding among the great variety of potential activities is their potential benefit for task accomplishment (e.g., "does this activity help me in accomplishing my tasks?").

Similarly to task goals, task orientation encompasses both a cognitive and a motivational component (cf., Frese and Zapf, 1994; Hacker, 1994; Hacker, 1998). The cognitive component corresponds to an adequate mental representation of present and potential future tasks. Thus, it is part of an individual's work-related mental model. For example, an individual with a high cognitive task orientation knows what his or her present tasks are and in which direction they might develop in the future. In contrast, an individual with a low cognitive task orientation does not have a clear representation of his or her tasks and how they might look like in the future. To have a high cognitive task orientation implies to have a good representation of the goal to be achieved and a good understanding of the steps necessary to achieve this goal.

The motivational component refers to an individual's commitment to reach the task goal. For example, an individual with a high motivational task orientation experiences task accomplishment and goal attainment as highly important and is willing to put effort on activities, which lead to goal attainment. An individual low on motivational task orientation does not care much about goal attainment or might even avoid putting effort in the task accomplishment process (Nicholls, 1984). The motivational task orientation concept shows substantial overlap with related concepts such as mastery orientation or learning goal orientation (Dweck and Leggett, 1988).

Conceptually, the cognitive and motivational components of task orientation are two distinct dimensions. An individual high on the cognitive component may or may not be high on the motivational component. For example, one can imagine an individual who possesses a high cognitive task orientation and at the same time experiences a low motivational task orientation. This implies that this person knows exactly what his or her tasks are, but is not motivated to accomplish them. In contrast an individual may be high on the motivational component but low on the cognitive component. This might particularly be the case for newcomers in an organisation (Miller and Jablin, 1991) or for individuals working in modern work settings characterized by vague task descriptions that require a broad role orientation from the employees (Parker, Wall, and Jackson, 1997). In these settings, individuals might be very motivated to do their work – but they do not know exactly what their tasks are, in which direction these tasks are going to develop in the future, and what they should do to accomplish their present and future tasks.

If tasks are accomplished in a teamwork setting, matters are even more complex. Research on transactive memory (Wegner, 1987) has shown that individuals who work together in a team have a shared representation of the tasks of their fellow team members and of the respective knowledge they need for accomplishing their tasks (Liang, Moreland, and Argote, 1995; Moreland, 1999). Thus, for performing well in a teamwork setting it is not enough that individuals are high on the individual aspect of cognitive task orientation (possessing a good mental representation of their own present and future tasks) but they are also high on the team-related aspect of cognitive task orientation (i.e., possessing a good mental representation of their fellow team members' present and future tasks).

Intensity and Selectivity of Knowledge Management Behaviour

For understanding individual knowledge management and its potential relationship to information overload, it is crucial to differentiate between intensity of knowledge management and selectivity in knowledge management. Intensity refers to the pure amount of knowledge management activities and mainly focuses on the issue of quantity. An individual high on the intensity of knowledge management will put great effort in knowledge search, knowledge storage and knowledge distribution without necessarily prioritising specific pieces of information or knowledge areas. An individual low on the intensity of knowledge management will be reluctant in searching for knowledge, will not store it, and will not inform his or her co-workers about issues that might be helpful or necessary for their work.

Selectivity in knowledge management refers to choice behaviour within knowledge management. Selectivity implies prioritisation and selection processes during knowledge search, storage, and knowledge distribution. It does not refer to the sheer amount of knowledge to be acquired, stored, and distributed, but to its perceived potential value for task accomplishment. An individual with a high selectivity in knowledge management will look for specific contents of knowledge, will deliberately decide whether it is worthwhile to keep a 'tangible' reference to this knowledge (e.g., a written note, a computer file, a hyperlink), and will inform co-workers only about information and knowledge areas anticipated to be helpful for them. In addition, one can differentiate between an individual and a team-related aspect of selectivity in knowledge management. The individual aspect mainly addresses knowledge search and storage, the team-related aspect is relevant for knowledge storage and distribution.

One can assume that intensity and selectivity in knowledge management show a weak positive relationship because a certain amount of intensity in knowledge management is a prerequisite for being selective. Without any knowledge management activity one cannot be selective. However, one can also imagine instances in which intensity and selectivity do not co-vary. In extreme cases, an individual high on the intensity of knowledge management will look for as much knowledge as possible, always retain a 'tangible' reference to this knowledge, and pass it on to his or her co-workers, regardless whether this knowledge is potentially useful for the person him- or herself or this person's co-workers.

A Model on Individual Knowledge Management Behaviour

Figure 15.1 shows the proposed model on the relationship between task orientation, knowledge management behaviour and information overload. The model differentiates between cognitive and motivational task orientation as well as between intensity and selectivity of knowledge management. The basic assumption is that cognitive and motivational task orientation are related to individual knowledge management behaviour with the two task orientation components showing differential pattern with intensity and selectivity of knowledge management. Specifically, the model assumes that a high motivational task orientation has a positive effect on intensity of knowledge management. In the case of a high motivational task orientation, an individual will be highly involved in actively searching for knowledge, distributing it to co-workers, and storing it for potential future use. In case of a low motivational task orientation, an individual will be less willing to search for knowledge, distribute, or store it.

Proposition 1. A high motivational task orientation is positively related to a high intensity of knowledge search, knowledge distribution, and knowledge storage.

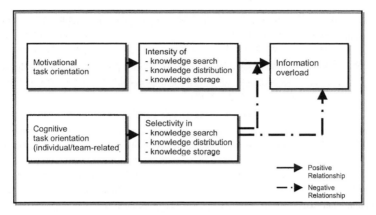

Figure 15.1: Task orientation, knowledge management and information overload

The cognitive task orientation component plays an important role for the selectivity in knowledge management, i.e. prioritisation and selection processes. An individual high on the individual aspect of cognitive task orientation knows what his or her work tasks and related goals are and how they will look like in the future. This individual will proceed selectively when searching and storing information and knowledge. In the optimal case this person will only look for and store information that he or she needs for accomplishing present and anticipated future tasks. An individual however, who is low on the individual aspect of cognitive task orientation and has no clear representation of his or her present and future tasks, can not be selective in searching and storing information. This person does not know which information or knowledge is important and which is not. Therefore, this person will show only a low selectivity in knowledge management behaviour. Particularly, if this person is high on motivational task orientation, he or she will collect and store as much information and knowledge as possible.

Proposition 2. A high cognitive task orientation (individual aspect) is positively related to a high selectivity of knowledge search and knowledge storage.

An individual who is high on the team-related aspect of cognitive task orientation has a good mental representation of the other team members' tasks. Therefore, this person will be selective when storing and distributing information and knowledge. This person knows who in the team needs which information and is therefore able to distribute information selectively to other team members. An individual however, who does not have a good mental representation of his or her fellow team members' tasks will distribute all the available information to all the other team members.

Proposition 3. A high cognitive task orientation (team-related aspect) is positively related to a high selectivity of knowledge storage and knowledge distribution.

Intensity and selectivity of knowledge management behaviour have differential effects on information overload. The model assumes that a high intensity of knowledge management behaviour results in information overload. Specifically, a high intensity of knowledge search and storage will cause information overload in the person him- or herself, a high intensity of knowledge distribution will cause information overload in this person's co-workers.

Proposition 4. A high intensity of knowledge search and knowledge storage is positively related to information overload.

Proposition 5. A high intensity of knowledge storage and knowledge distribution is positively related to co-workers' information overload.

Selectivity of knowledge search however, will show a negative relationship with information overload. If a person prioritises information and knowledge and is selective in storing and distributing it, it is less likely that information overload will occur. In cases of low selectivity, information overload will increase. More specifically, selectivity in knowledge search and storage helps a person to prevent information overload in him- or herself while selectivity in knowledge distribution is important for preventing information overload in his or her co-workers.

Proposition 6. A high selectivity of knowledge search and knowledge storage is negatively related to information overload.

Proposition 7. A high selectivity of knowledge storage and knowledge distribution is negatively related to co-workers' information overload.

Selectivity in knowledge management behaviour will be particularly useful in instances of high intensity of knowledge management behaviour. On the one hand this implies that in the case of a high intensity of knowledge search and knowledge storage, an individual needs to proceed selectively to prevent information overload. Framed differently, high cognitive task orientation buffers the negative effects of intensive knowledge search and knowledge storage on information overload. On the other hand, this implies that information overload will be particularly high in the case of high intensity and low selectivity in knowledge management behaviour, i.e. when an individual is very active and highly motivated to search, store and distribute information but does not make any informed choices about with pieces of information and knowledge are relevant for task accomplishment.

Proposition 8. A high selectivity of knowledge search, knowledge distribution, and knowledge storage moderates the unfavourable effects of a high intensity of knowledge search, knowledge storage and knowledge distribution on information overload.

Boundary Conditions

The proposed relationships between task orientation, knowledge management behaviour, and information overload will not hold for all circumstances. Among others, such boundary conditions refer to an individual's skills and knowledge, characteristics of the team, and the task to be accomplished.

A high motivational task orientation might not be a sufficient condition for a high intensity of knowledge management behaviour. For example, a high motivational task orientation might only have no effect on knowledge search unless a number of conditions are met: First, an individual must know relevant knowledge sources and must possess the necessary skills for using these knowledge sources. Second, an individual must have access to these knowledge sources. Third, an individual must have the opportunity to engage in knowledge search. For example, one can imagine, that high time pressure and tough deadlines for completing tasks may impede extensive knowledge search, although when motivational task orientation is high.

Similarly, a high motivational task orientation might not be enough for distributing information and knowledge to fellow team-members or other co-workers. First, if task interdependence is low, individuals might see no immediate necessity to distribute knowledge and information. Second, if individuals expect negative consequences from sharing knowledge and information they will be reluctant in distributing knowledge and information. Among others, such anticipated negative consequences might include expected loss of power and the fear the one's career opportunities will deteriorate. Taken together, expected benefit from knowledge management is a necessary condition for the effects of motivational task orientation on knowledge management.

Also high cognitive task orientation might not always lead to a high selectivity of knowledge management behaviour. I assume that the relationship between cognitive task orientation and selectivity in knowledge management behaviour is contingent on the phase of the task accomplishment process. For example, in early phases of task accomplishment also an individual with high cognitive task orientation may engage in extensive knowledge search, storage, and distribution. For example, imagine a researcher who has a good mental representation of the research task but nevertheless extensively searches for information that might help in the research process. However, later in the task accomplishment process, high cognitive task orientation will be particularly helpful in sorting out irrelevant pieces of information.

Moreover, a high intensity of knowledge management behaviour will not necessarily result in information overload. The model spelt out that selectivity in knowledge management might play a major role. Moreover, also an individual's cognitive representation of the field might be important. For example, an individual with a highly developed and well-integrated mental model of his or her field of expertise will experience less information overload, also when showing a high intensity of knowledge management behaviour.

PRELIMINARY EMPIRICAL RESULTS

An empirical study in information technology firms tested parts of the model presented in this chapter (Sonnentag, in prep.). Specifically, this study examined the relationships between individual task orientation, individual knowledge management activities, and information overload (Propositions 1, 2, 4, and 6).

The study sample comprised 200 employees of 40 small and medium-sized information technology companies located in Southern Germany and Switzerland. Cognitive task orientation was assessed with measures of goal clarity and process clarity developed by Sawyer (1992). Motivational task orientation was operationalised as learning goal orientation (VandeWalle, 1997). Knowledge management behaviour and information overload were measured with scales developed by Akli and Sonnentag (2001). Specifically, intensity of knowledge management was assessed by two scales (a) active search for knowledge and (b) distribution of knowledge. Selectivity of knowledge management was measured with a scale focusing on prioritisation of important knowledge. Data were analysed by following a hierarchical regression analysis approach. In these analyses a number of background variables including job tenure and task characteristics were controlled for.

Analysis showed that motivational task orientation was a significant predictor of active search for knowledge and for knowledge distribution. Cognitive task orientation predicted prioritisation of important knowledge. In addition to these hypothesized relationships, also the cognitive task orientation variables predicted distribution of knowledge and motivational task orientation predicted prioritisation of important knowledge. Cognitive task orientation variables however were no significant predictors of active search for knowledge. Analyses further revealed a negative relationship between prioritisation of important knowledge and information overload which was marginally significant. Additional inspection of the data suggested that particularly individuals with high process clarity suffered less from information overload.

Taken together, this study supported parts of the proposed model. Most important, individual task orientation was related to intensity and selectivity of knowledge management behaviour. This implies that individuals with a high motivational and cognitive task orientation were more involved in knowledge management behaviours than individuals with low task orientations. This suggests that an individual's orientation towards his or her work tasks is crucial for knowledge management. The study only partially supported the predicted differential effects of motivational and cognitive task orientation. Motivational task orientation was not only related to intensity of knowledge management, but also to selectivity. Cognitive task orientation was not only related to selectivity, but also to one of the intensity aspects of knowledge management. Thus, the motivational aspect of task orientation is relevant for all studied aspect of knowledge management behaviour, whereas the cognitive aspect comes into play in prioritising important knowledge and distributing knowledge. The effects of knowledge management on information overload were less clear-cut. However, the analysis suggests that information overload is related to a lack in cognitive task orientation (primarily process clarity) and low selectivity in knowledge management.

Of course, this study has some limitations, primarily its cross-sectional nature and its focus on self-report measures. Therefore, causal inferences cannot be drawn. For example, it can not be ruled out that individuals partly react to information overload with specific knowledge management behaviours. In the future, longitudinal studies are needed that address the issue of causality.

CONCLUSIONS

The model proposed in this chapter suggests that individual task orientation is crucial for individual knowledge management behaviour. More specifically, the model assumes that motivational task orientation is most relevant for the intensity of knowledge management, whereas cognitive task orientation is most relevant for the selectivity of knowledge management. Only if an intensive knowledge management behaviour is accompanied by a high selectivity, information overload can be reduced or avoided.

The empirical study showed that individual task orientation is related to individual knowledge management behaviour. The more an individual focuses on the task to be accomplished and on related goals, the more this individual engages in knowledge management and prioritises important knowledge and information. Motivational task orientation is highly relevant, both for intensity *and* selectivity of knowledge management.

If these findings are supported in longitudinal causal analysis, implications for practice are obvious: to enhance individuals' intensity and selectivity of knowledge management, individuals' task orientation has to be addressed. Organisations may try to select individuals who are highly task-oriented. In addition, re-framing tasks as learning tasks that require knowledge search activities might be useful. Furthermore, it will be necessary to be clear and specific about the goals to be attained and the ways to achieve them.

Thus, for engaging in selective knowledge management, a clear representation of one's present and future work tasks is a prerequisite. However, individuals and organisations might face a dilemma here: for promoting pro-activity and flexibility organisations increasingly avoid precise task descriptions and encourage broad role orientations (Parker *et al.*, 1997). This however, leaves individuals with vague ideas about their tasks and responsibilities. Selectivity in knowledge management becomes extremely difficult and information overload might increase.

Furthermore, the discussion of this model's boundary conditions points to additionally important prerequisites of knowledge management. Individuals will only engage in knowledge management behaviours, if they have the knowledge, skills, and opportunities to do so. They will only invest in the distribution and storage of knowledge if they see any benefit of it. If they however can accomplish their tasks without any input from their co-workers (and vice versa) or expect negative consequences from knowledge sharing they will withhold from any greater effort in knowledge management.

Clearly more empirical research is needed which addresses the prerequisites and consequences of personalized, individual knowledge management behaviour. An organisation's investment in knowledge management, including codifying efforts, will only pay off if individuals are able and willing to use these knowledge sources and knowledge management methodologies in the processes of accomplishing their work tasks.

ACKNOWLEDGEMENT

The empirical study reported in this chapter was supported by a research grant from the University of Konstanz (Germany), which is gratefully acknowledged. I thank the members of the work and organisational psychology group at the University of Konstanz for helpful discussions and the participants and organisers of the 19th NetWORK meeting, particularly Erik Andriessen and Babette Fahlbruch for valuable comments on an earlier version of this paper.

REFERENCES

Akli, H., and S. Sonnentag (2001). *Knowledge management in the work context: Development of an instrument for assessing individual knowledge management*. Paper presented at the 10th European Congress of Work and Organizational Psychology, May 16-19, 2001, Prague (Czech Republic).

Davenport, T. H., and L. Prusak (1998). *Working knowledge: How organizations manage what they know*. Harvard Business School Press, Boston, MA.

DeLong, D., and P. Seemann, (2000). Confronting conceptual confusion and conflict in knowledge management. *Organizational Dynamics*, **29**, 33-44.

Dweck, C. S., and E. L. Leggett (1988). A social-cognitive approach to motivation and personality. *Psychological Review*, **95**, 256-273.

Fleishman, E. A. (1953). The description of supervisory behavior. *Personnel Psychology*, **37**, 1-6.

Frese, M., and D. Zapf (1994). Action as the core of work psychology: A german approach. In: *Handbook of industrial and organizational psychology* (H. C. Triandis, M. D. Dunnette, and L. M. Hough, eds.), Vol. 4, pp. 271-340. CA: Consulting Psychologists Press, Palo Alto.

Grant, R. M. (1996). Prospering in dynamically-competitive environments: Organizational capability as knowledge integration. *Organization Science*, **7**, 375-388.

Hacker, W. (1994). Action regulation theory and occupational psychology: Review of German empirical reserach since 1987. *German Journal of Psychology*, **18**, 91-120.

Hacker, W. (1998). *Allgemeine Arbeitspsychologie: Psychische Regulation von Arbeitstätigkeiten*. Huber, Bern.

Hansen, M. T., N. Nohria, and T. Tierney, (1999). What's your strategy for managing knowledge? *Harvard Business Review*, **77**(2 (March-April)), 106-116.

Liang, D. W., R. L. Moreland, and L. Argote (1995). Group versus individual training and group performance: The mediating role of transactive memory. *Personality and Social Psychology Bulletin*, **21**, 384-393.

Miller, V. D., and F. M. Jablin (1991). Information seeking during organizational entry: Influences, tactics, and a model of the process. *Academy of Management Review*, **16**, 92-120.

Moreland, R. L. (1999). Transactive memory: Learning who knows what in work groups and organizations. In: *Shared cognition in organizations: The management of knowledge* (L. L. Thompson, J. M. Levine, and D. M. Messick, eds.), pp. 3-31. Lawrence Erlbaum, Mahwah, NJ:.

Nicholls, J. G. (1984). Achievement motivation: Conceptions of ability, subjective experience, task choice, and performance. *Psychological Review*, **91**, 328-346.

Nonaka, I. (1994). A dynamic theory of organizational knowledge creation. *Organization Science*, **5**, 14-37.

Parker, S. K., T. D. Wall, T. D., and P. R. Jackson (1997). "That's not my job": Developing flexible employee work orientations. *Academy of Management Journal*, **40**, 899-929.

Pazy, A. (1994). Cognitive schemata of professional obsolescence. *Human Relations*, **47**, 1167-1199.

Probst, G., S. Raub, S., and K. Romhardt (2000). *Managing knowledge: Buildings blocks for success*. Wiley, Chichester.

Sawyer, J. E. (1992). Goal and process clarity: Specification of multiple constructs of role ambiguity and a structural equation model of their antecedents and consequences. *Journal of Applied Psychology*, **77**, 130-142.

Sonnentag, S. (in prep.). *Individual knowledge management in information technology firms*. Working paper. Technical University of Braunschweig.

Sparrow, P. R., and K. Daniels (1999). Human resource management and the virtual organization: Mapping the future research issues. In: *Trends in organizational behavior* (C. L. Cooper and D. M. Rousseau, eds.), Vol. 6, pp. 45-61. Wiley, Chichester.

Tesluk, P., D. Hofmann, and N. Quigley (2002). Integrating the linkages between organizational culture and individual outcomes at work. In: *Psychological management of individual performance* (S. Sonnentag, ed.), pp. 441-469. Wiley, Chichester.

Ulich, E. (1991). *Arbeitspsychologie*. Verlag der Fachvereine, Zürich..

VandeWalle, D. (1997). Development and validation of a work domain goal orientation instrument. *Educational and Psychological Measurement*, **57**, 995-1015.

Wegner, D. M. (1987). Transactive memory: A contemporary analysis of the group mind. In: Theories of group behavior (B. Mullen and G. R. Goethals, eds.), pp. 185-208. Springer, New York.

16

THE COGNITIVE AND ORGANISATIONAL PROCESSES OF INNOVATION: A KNOWLEDGE MANAGEMENT PERSPECTIVE

John Hurley[1]

INTRODUCTION

This chapter outlines the importance of knowledge management in knowledge intensive organisations. A theoretical cognitive-organisational framework describing the process of knowledge management is proposed, with an emphasis on knowledge creation and innovation. This cognitively linked dynamic framework comprises a number of disparate but related concepts, where the proposed function of knowledge management is the co-ordination of these apparently incongruent concepts. Such concepts as organisational learning, product development, problem identification and solution, and technology are perceived as pivotal in the development of a knowledge-creating organisation.

The first section of this chapter describes knowledge management and shows its links with innovation and knowledge creation. The next section outlines the changing forms of organisation needed for knowledge development and creation to occur. The third section draws together a number of the emerging themes in an attempt to suggest a theoretical cognitive-

---

[1] Dublin City University, Ireland

organisational framework. The chapter concludes by describing ways of testing the cognitive-organisational theory suggested.

THE IMPORTANCE OF KNOWLEDGE MANAGEMENT IN RELATION TO THE KNOWLEDGE CREATING ORGANISATION

In a world characterised by a continuing desire to improve the standard of living, competitiveness is central to the success of this desire, and applies in most fields. It extends from manufacturing industry, to service providers, to finance, to software development and even to universities and other institutes of higher education. If an organisation wishes to remain or become competitive, it must constantly innovate to improve its products or services. If these products or services lag behind those of other providers then competitive advantage is lost and consumers/clients will take their business elsewhere. Hence competitive innovation is strongly linked to survival.

The organisation that continues to make cassette recorders long after CD's are in general demand, is one which cannot adapt quickly, and may not survive another rapid technological shift. Olivetti's failure to recognise and adapt to the technological advances of the word processor almost ruined the typewriter giant. Such organisations are often created purposely for stability of manufacturing, to control quality, and to resist change, and are typical of bureaucracies. Organisations which innovate, learn, and adapt quickly, are of a very different kind, and are now based on knowledge and learning and on an adaptable, flexible form of management.

When a product or service does not exist, and the aim is to bring it into existence, as in product development, then we need to know very clearly the nature of the group that will do that best. It is unlikely that knowledge will be elicited in a pyramidal hierarchy of more than three levels; knowledge elicitation it seems, is more is more likely to occur in a collegial grouping, designed for sharing information, and collaborating in the development of that information and the ideas that emerge from it. We can think of the knowledge requirement in organisations as extending across an organisational type continuum as illustrated in figure 16.1:

	knowledge complexity →		
Nature of work or organisation:	*Routine production or service*	*Innovative work or organisation*	*Scientific discovery*
Level of knowledge required:	Low	Moderate to high	Very high
Type of development needed:	Precise training	Elaborate developmental exercises aimed at building trust and collaboration	Great freedom and interaction with gifted colleagues

Figure 16.1: Continuum of organisational type.

This continuum of knowledge requirement ranges from routine production, through innovation to discovery. At the upper extreme of the continuum between routine production, and high level innovation, lies the person who makes discoveries Hence, we cannot speak of knowledge management in general, and must be careful to direct our conclusions to the appropriate type of organisation, as different organisations require different types of knowledge management. This chapter refers to innovative organisations where new knowledge is produced and applied.

There is growing evidence from recent research that when the knowledge requirements are different, as illustrated in figure 16.1, then the organisational requirements also become different. In a study among science Nobel Laureates, Hurley (1997) has shown that eminent scientists value freedom above everything else if their science is to succeed. Table 16.1 below, reports the results of the questionnaire administered to 14 Nobel Laureates, which investigated the importance of the role of freedom in scientific discovery. It is clear that this group of eminent scientists regards freedom as of absolute importance, without which their minds cannot construct a new reality, which is necessary for discovery and invention.

These eminent scientists are an extreme case of knowledge management, at the far end of the continuum, and greatly removed from routine workers. But they illustrate the way organisations must change if they are to accommodate knowledge within their work, and not just work and manage according to previously developed systems, following instruction manuals, and other methods of prescribed systematic control.

Table 16.1:   Nobel Laureates Mean Responses Concerning the Importance of Certain Aspects of Work for Scientific Discovery.

Q. No.	Question	Mean	Std Error of mean
9	Freedom of thought	4.00	0.0
1	Being able to choose what I work	3.92	.08
8	Independence of thought	3.92	.08
3	Freedom to use my own judgement	3.83	.11
2	Free to choose my own problems	3.73	.14
6	Working to a defined research plan	2.67	.22
5	Pressure from being supervised	2.18	.38
4	Working as part of a team	2.17	.30
7	Pressure from peers	1.83	.21

Note: scale 1=very unimportant, 4=very important (Hurley, 1997, p.4).

The freedom the Nobel Laureates speak of is mostly a freedom of the mind, unaffected by social or organisational pressures. It is:

Not absolute freedom, and not endless time and boundless resources, but freedom above all to use ones own personality in pursuit of a scientific objective. Freedom to pursue hunches down possibly pointless avenues of exploration; freedom to theorise, experiment, accept, reject, according to the principal investigators own judgement, with no other interference. Considerable organisational resources are needed to allow this freedom, and give the principal investigator time to reach a satisfactory conclusion Hurley (1997, p.4).

This freedom need is a strong component of the collaborative, idea-generating practices, so important to innovation. There is great need for this freedom when knowledge is being created or developed, whereas where knowledge is being applied, systems can be created which ensure standard products and services, and which necessarily limit individual freedom. Similar findings on the importance of freedom to innovation and creativity have been made by Andrews and Gordon (1970), and Amabile (1988).

By contrast with the Hurley study, Mouly and Sankaran (1998), conducted an ethnographic study in a large scientific institute where productivity was not high, with multiple laboratories, and concluded that if laboratories provide poor resources, excessively routine work, inadequate leadership and strict controls, the probability of effective research is low. Though this may appear to be obvious, it has not been shown in a scientific methodology to have such an important influence on outcomes, until the above study was carried out. Taken together both the Hurley study and the Mouly and Sankaran one, demonstrate the strong organisational and group connection with scientific effectiveness.

If we are to structure our innovative or discovery oriented organisations in ways that meet the needs of creative and innovative people, we must make important changes in the nature of societal and organisational life. Much of the productive enterprise of the future is likely to move away from manufacturing and towards advanced services such as basic or applied research, software development, product innovation, and discovery. It seems likely that this form of work is going to employ many more people than production enterprises, which will probably move to the Pacific Rim countries, and will in any case be machine and robot supported. We have seen this happen in many industries before, for example with cotton and other materials producing sectors, that shifted to developing countries, while the technology and fabric design of this industry remained in the developed world.

Organisations will need to adapt their processes and practices in order to accommodate this very different kind of dynamic created by rapidly changing demand, and greater competitive drive. This is what knowledge management is about.

CHANGING ORGANISATIONAL PROCESSES

Hurley (1990) drew attention to the 'collaborative imperative' of technology, which has forced the introduction of collaborative practices at work as a result of the requirements of technological change faster than thirty years of legislation had done.

The reasons for this rapid move to collaborative systems lie fundamentally in the importance of knowledge to modern economies, in particular in the way it adds value to goods and services and makes them more innovative and competitive. Collaborative relationships at work, (as distinct from adversarial ones) make it easier for tacit knowledge to be elicited, and made explicit. According to Polanyi (1958) knowledge about the object or phenomenon that is in focus is focal knowledge. Knowledge that is used as a tool to handle or improve what is in focus is tacit knowledge. When tacit knowledge is expressed through language, it becomes explicit knowledge. Only then can it be focused for reflection and development. It is only when it is explicit that innovation or discovery can be achieved, as creative ideas are often latent and need to be brought to the surface. Collaborative processes such as those described by Nonaka and Takeuchi (1995) are essential to bring out and make explicit the tacit knowledge that may otherwise remain untapped. The processes they describe include: the development of teams; socialization at the work team level; and developing a common language. By these systems they ensure that knowledge of a tacit kind is elicited and shared in the group, so that a common level of knowledge is developed.

In respect to the importance of knowledge within the organisation, the change over the last century is quite phenomenal. In Frederick Taylor's time, and using the technologies then in use, knowledge was assumed to reside at the top of the organisation, and the actual production took place at the bottom of the hierarchy. Gradually, but at an accelerated pace in the last five years, service organisations such as those in finance, advisory work, and internet activity have found that this hierarchical arrangement has had to be almost reversed. Many of the key areas of knowledge now exist only at the bottom of the organisation, with management uninformed on specific detail. For knowledge-based organisations this requires de-layering to take place within the hierarchy, in order to facilitate the exchange of tacit knowledge, and its' development into explicit systems. Thompson and Warhurst (1998) see the process as follows:

> "Put another way, the old vertical division of labour will be replaced by horizontal co-ordination. This is driven by the nature of knowledge work itself, which is essentially concerned with problem solving, problem identifying and strategic brokering between the two processes" (p.2)

It seems likely that the knowledge organisation and its' relationship with individuals involved in them will change because of the new emphasis on innovation and learning.

Tovstiga (1999) makes a related point:

> ".... increasingly, organisations also understand that their knowledge processes are inextricably linked to the organisation' s internal context - its internal management practices, learning culture and knowledge base. This has particularly been found to be the case with the tacit form of knowledge - that highly intangible and essential part of 'knowing how' and 'knowing why' on which technological innovation is primarily dependent."(p.732)

We need to know the nature of the process by which this innovation is introduced, and its cognitive linkages and reference points. An attempt will be made here to delineate these processes and linkages. It must be regarded as tentative and exploratory, and by no means as certain as the description of it might suggest.

The organisational processes and stages related to innovation and competitiveness can be conceived as a continuous process along the lines outlined in figure 16.2 below.

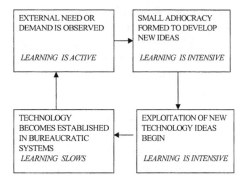

Figure 16.2:   The innovation process and the stages of organisational learning in the Western world, Hurley (2000, p. 6)

Hurley (2000) sees knowledge and learning as central to the creation of further knowledge creation, and thence to innovation. According to Hurley, learning intensity increases as ideas (knowledge) develop, and as ideas become established learning begins to slow. Knowledge is therefore created, developed and applied, but this process is a continuous one.

Knowledge and learning are linked with innovation and change in a way, which points to the more organic structure of organisations of the future. Nonaka and Takeuchi (1995) see the linkages as follows in figure 16.3.

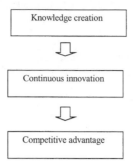

Figure 16.3:   The process of knowledge development in Japanese companies (Nonaka and Takeuchi, 1995, p.6)

The phrase 'continuous improvement' or innovation relates to the Japanese philosophy of 'Kaizen'. Knowledge creation is continuous for certain Japanese firms, whereas in the Western world knowledge creation is a catastrophic process arising from growing dissatisfaction with existing knowledge, or products or services, as illustrated in figure 16.2. According to Nonaka and Takeuchi (op.cit.):

> "Japanese companies, however, have a very different understanding of knowledge. They recognize that the knowledge expressed in words and numbers represents only the tip of the iceberg. They view knowledge as being primarily "tacit"-something not easily visible and expressible. Tacit knowledge is highly personal and hard to formalize, making it difficult to communicate or to share with others. Subjective insights, intuitions, and hunches fall into this category of knowledge. Furthermore, tacit knowledge is deeply rooted in an individual's action and experience, as well as in the ideals, values, or emotions he or she embraces" (p.8).

In their view, the West with its emphasis on information and data, and on that which is explicit, miss perhaps the most important resource available within organisations, the tacit knowledge lying often untapped within the individual.

How does the organisation encourage this to happen? Tovstiga (1999) describes the complexity of interactions within the knowledge based organisation as comprising three main elements which they illustrate in figure 16.4. The need to manage the processes of knowledge development are illustrated here.

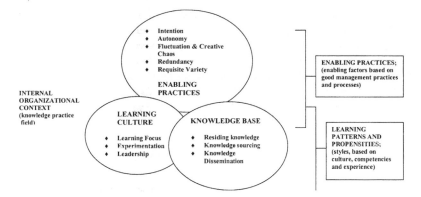

Figure 16.4: Dimensions of the internal organisational context. (Tovstiga 1999, p.736)

In this diagram, Tovstiga (1999) brings together the five enabling practices of Nonaka and Takeuchi, with the organisations knowledge base, and with its learning culture. Tovstiga's view is that all these factors have to be consistently present for effective knowledge management to take place.

Apart from the 'enabling practices' which are of overriding importance, the knowledge based organisation will pay attention to the development of its knowledge base, ensuring that it makes explicit and shares as much of its residing tacit (non-expressed) knowledge, and that its sources of new knowledge are very efficient, along with its systems for the dissemination of this knowledge. The organisation will have a strong learning focus, will be oriented to experimentation, and its leadership will support this focus on a continuing basis.

Tovstiga (1999) supports the positions of Nonaka and Takeuchi (1995) and Hurley (2000) that knowledge is core to competitive advantage. Tovstiga sees learning as creating knowledge, which then fuels innovation. However, unlike the proposed cognitive-organisational model presented in this chapter (figure 16.5), Tovstiga states that innovation breeds change and accelerates learning, whereas in the proposed model there is a slowing down of learning as the new technology becomes established in bureaucratic systems. Hurley's model describes the actual situation with regard to the emergence of new products in many manufacturing organisations, typically in western industry. The Nonaka and Takeuchi model describes how knowledge can be harnessed within the workforce to bring about new product development in a more organic way typical of the Japanese 'Kaizen' or continuous improvement approach, taking into account gradually changing consumer requirements as well as the internal expertise of the plant worker. Essentially, what both these models illustrate is that because of the increasingly rapid rate of technological change and market innovation, knowledge is at the very centre of organisations competitive advantage. It follows then that creating the right conditions within the organisation to facilitate the emergence of new knowledge, is an important priority for those managers attempting to gain competitive advantage through knowledge. The organisational nature of this facilitation process is the subject of the next section.

A POSSIBLE COGNITIVE – ORGANISATIONAL FRAMEWORK

What has changed in the last twenty or so years and made knowledge management so much more important, is that the advances in technology and much greater computing power and information retrieval have become available, and this has meant that the conceptualisation of Polanyi (1962) regarding the international work of science, has become a reality.

Polanyi (1962) saw that scientists throughout the world behave as if they were working on the joint development of a huge jigsaw puzzle, working separately, but in a coordinated way has to a great extent become a reality. The Internet has joined those developing knowledge into a closer knit world community; bibliographic data bases have become widely available to the researcher. The world wide web has made more discussion of ideas possible without the necessity to travel. We have seen that knowledge management is important in relation to innovation, and thereby to competitiveness. Tovstiga's (1999) dimensions of the organisation (figure 16.4), bring together the theories of Nonaka and Takeuchi (1995) and Hurley (2000) highlighting the importance of the learning culture, knowledge base and enabling practices in knowledge management. The proposed cognitive-organisational framework below builds on the work of these authors and is expanded and developed into a broader theoretical model. (See figure 16.5).

The cognitive-organisational framework upon which our organisational processes are based needs to be defined and developed in a precise and usable way. In order to achieve this it becomes necessary to explain the reasons for the different approaches to knowledge management, knowledge creation and innovation, taken by the East and West. The two cultures are based on different philosophical premises, which are at times complementary, overlapping and disparate. Broadly speaking Western approaches to innovation are based on the scientific method. In the East on the other hand a non-empirical philosophy has traditionally been the basis of their approach to innovation and knowledge creation.

Nonaka and Takeuchi (1995) have the view that the West has ignored the rich resources among the personnel in their own organisations. Their view of the process of innovation and knowledge creation is based in 'Kaizen'. Traditionally, in the West we have approached the innovation process from several psychological and scientific The scientific approach is based on empirical testability. Though flawed and subject to revision, it is a most important objective mode of pursuing knowledge. The scientific method aims to produce knowledge of the world by the establishment of generalisations governing the behaviour of the world (Chalmers, 1990). Taylorism and Fordism were based on this.

A popular understanding of the process of innovation is the individual trait approach, the emphasis being on pre-dispositions towards behaviour, in this case innovation. This does not necessarily exclude environmental influences, but instead establishes a reaction range of possible innovative behaviours.

Behaviourism, ignores internal cognitions but rewards the observed behaviour to be encouraged (innovation), and extinguishes the behaviours to be eliminated (non-innovation). According to the phenomenological approach, there is no objective reality only our interpretation of that reality. Proponents of this perspective include Maslow and Rogers. Phenomenology proposes that we construct our own interpretations of the world and live in a shared reality with others. This approach is not common in organisations, and is closer to the Japanese approach, but uses 'techniques' like non-directive counselling to elicit knowledge.
The social cognitive approach emphasises how our thoughts interpret social interaction, which reflects the social world and includes formal logic. It also looks at how we regulate our behaviour, in this instance innovative behaviour

These theories, essentially emanating from the scientific model , have a commonality. They all have concomitant techniques, which can be applied to organisations generally, to achieve a behavioural goal, i.e. innovation.

The Western approach is scientific, incorporating the psychological theories, with phenomenology being the closest to the eastern approach. These psychological theories highlight the 'individualistic' approach to innovation taken by the West; each of these approaches has a technique to elicit behaviour or knowledge and have been used in, or at least influenced, management practices. These theories ignore the importance of 'tacit' knowledge, preferring to see knowledge as object, to be codified and used. Knowledge creation according to Nonaka and Takeuchi requires a different philosophy and approach, which can incorporate knowledge that is not always objective or codified.

Polanyi (1966) at a philosophical level and Nonaka and Takeuchi (1995) in a more applied way, have both drawn attention to the central importance of eliciting 'tacit' knowledge. Neither gives entirely satisfactory descriptions as to how this is done, nor what the psychological or group mechanisms underlying the process are. Nonaka and Takeuchi refer to the group socialisation process, and criticise western organisations for being excessively influenced by rational models such as Taylorism. They suggest that there is a wealth of tacit knowledge that lies largely unknown in many organisations. In creating a possible cognitive-organisational framework for knowledge management in relation to innovation, it may be argued that it is this tacit knowledge and the elicitation and application of it that is the key to competitiveness and innovation as opposed to knowledge in general. Tacit knowledge may be seen as an fundamental aspect throughout the following cognitive-organisational framework.

Eleven dimensions of knowledge management , (each of which will be discussed later) are proposed in the framework. Of these eleven dimensions, four are organisational, five involve group dynamics, and two are cognitive. Inevitably, there is overlap, for in this field definitions are few, and borders and interpretations many.

The organisational aspects are:
- learning culture
- requisite knowledge
- requisite technical systems
- equitable reward system

The group dynamics variables are:
- knowledge sharing
- double-loop learning
- active idea-generation and processing
- smart foraging
- avoidance of groupthink

The cognitive aspects highlighted are:
- problem finding
- problem solving

Earlier, in this chapter the importance of the learning culture (Tovstiga, 1999, p. 736; Hurley, 2000, p.6), and requisite knowledge were outlined as the organisational bases for this cognitive framework. These organisational dimensions are now coupled with the group dynamic aspects of knowledge sharing and development, and with the cognitive aspects of problem finding and problem solving, to form the cognitive-organisational framework. In this model knowledge management is seen as having the function of creating the organisational climate and the group dynamics which will foster knowledge sharing and development

This framework forms the basis for a broader cognitive-organisational theory of knowledge management in relation to knowledge creation and innovation (figure 16.5). This theory encompasses all previously mentioned aspects of knowledge development and management and includes, in addition to the above-mentioned cognitive-organisational framework, aspects of group dynamics, which it is proposed, have a pivotal role in the success or failure of the development of new knowledge.

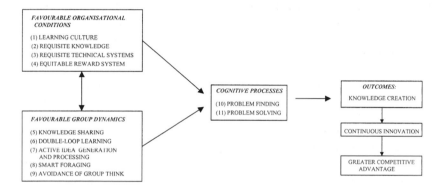

Figure 16.5: A cognitive-organisational model of the process of knowledge management

This model is put forward as a dynamic model of the key processes involved in knowledge management. Briefly, this model proposes that if there exists in an organisation, favourable conditions, then these should encourage the cognitive processes of problem finding and problem solving, which in turn will lead to successful outcomes i.e. knowledge creation. In addition, if there exists favourable group dynamics this should also foster the cognitive processes of problem finding and problem solving, leading to effective outcomes. The interaction of favourable organisational conditions and favourable group dynamics should have an even more positive effect on cognitive processes and consequent knowledge creation., than if the two conditions are taken separately. The cognitive processes act as mediators, in that increases in problem finding and problem solving lead to effective knowledge creation, however, the antecedent conditions both group and organisational need to be favourable. The model and the interactions suggested would need to be tested empirically, and an outline of a method of doing this is given at the end of this chapter.

The absence of such empirical testing in the area of knowledge management is perhaps partly because of the relative novelty of the field and partly because of the difficulty in quantifying its key concepts. To a great extent the importance of the above components to knowledge development are supported by the available literature. A brief summary of the immediately related literature is given here which leads into the programme for scientific testing which is the last section. The text is numbered to correspond to the numbers in the diagram.

Favourable organisational conditions

Learning culture
Senge (1990) defined a learning organisation as "an organisation that is continually expanding its capacity to create its future" (p. 14) He proposed five disciplines necessary for becoming a learning organisation, these were: personal mastery, mental models, shared vision, team learning and systems thinking. Senge's five disciplines have much in common with the Nonaka and Takeuchi (1995) approach, in that both emphasise sharing and knowledge elicitation. Organisational learning can also be seen as an application of Argyris and Schon's (1978) single loop (adaptive) and double-loop (generative) forms of learning (see chapter 2 in this book). Finerty (1997) in one study found that the establishment of a learning culture supported by powerful knowledge systems can transform the rate and quality of learning. Those factors, which Senge found to be central to the learning culture, not only helped accelerate the rate of learning and knowledge creation, but also provided a solid foundation for building meaning and motivation in the workplace.

The Finerty case study describes how Action Learning is helpful in releasing new knowledge in the workplace. Action Learning is defined by Revans (1983) as:
> "Action learning is a systematic process through which individuals learn by doing. At the basis of this type of learning is the thought that learning requires action and vice versa. This type of learning involves learners actively to reach the outcome of learning. Learning occurs as a result of learners pursuing solutions to problems".

In one study, Arad, Hanson and Schneider (1997) using factor analysis, developed a taxonomy of organisational characteristics which supported a six factor structure, these factors being: use of teams; formalization and standardisation; information sharing; decentralisation;

establishment size/ specialisation; and organisation size. Their taxonomy of organisational characteristics was developed in order to describe and classify jobs. They then proceeded to examine research related to innovation and their results suggest that certain characteristics within these factors, appear to foster innovation. In particular, the characteristics of innovation values, information sharing, decentralisation, goal-setting and the use of teams.

These results support a number of prior findings by Hurley (1997), Amabile (1988) Kanter (1999), and Pelz and Andrews (1966), including the importance of the team sharing information in order to develop knowledge.

To bring these findings together, we can see that for knowledge to develop within an organisation, it needs to be managed; part of that process of management is to ensure that the organisation culture permits and encourages new knowledge to develop, and is positive to experimentation and innovation.

Requisite information and knowledge
In addition to a learning culture, a resident information system and a knowledge system would be needed to support knowledge work. This would be needed to be put in place to ensure that the current information and knowledge would not be lost, for example if a colleague left the organisation. This could be modelled on an elaborate Management Information System, but designed around the ready availability of knowledge to each person. This would mean for example access to organisation records, online bibliographic databases etc. This idea would need to be developed to make it into a meaningful system.

Requisite technical systems
The existence in an organisation of the requisite technical systems to support the knowledge development process is essential. Depending on the nature of the technology in which innovation is being pursued, the technology required may need to be well in advance of existing state of the art technology, so that the technical base is sufficient to allow new developments. The development of instrumentation has always been essential for innovation. The development of the electron microscope made one series of advances possible; the development of scanning tunnelling microscopy another series. X-ray crystallography made peptide sequencing and human genome analysis possible. Hence for any innovation, continuing advances in technology are required to support it.

Existence of an equitable reward system

Organisational approaches to reward have tended to be divided into two situational areas. Where the individual's work does not greatly affect output as in mass production, then the rewards are fixed. When the individuals work affects output directly, then rewards are variable and linked to performance. This latter situation is illustrated by sales activity and by software development as examples. Reward systems have in some cases been used to shape behaviour, and it seems reasonable to expect that those involved with the growth of knowledge, should be rewarded if their work contributes to measurable improvements in effectiveness or competitiveness.

Of course, a sales person's output is easily measurable, in knowledge development, measurement is not so simple. To begin with it is a group activity, so the group might expect to be rewarded as a group. Secondly, though it is a group activity, the contribution of some individuals may be felt by most to be greater than others. Thirdly, the activity of the group frequently depends on the activities of yet other groups, systems, and technology within the larger organisation.

It is perhaps no accident then that the most widely used method of reward in the knowledge development area is in fact the granting of shares or share options in the organisation. Another widely used method is the system of rewarding suggestions known as 'Vorschlagswesen' used in Germany for about fifty years with considerable success (Nickel and Krems, 1998). (They found that "The dimensions of friendliness, activation, and co-operation were correlated with the suggestion rate"(p.27). This supports Nonaka and Takeuchi's view that tacit knowledge is elicited best by socialisation among colleagues. Nevertheless, this study also shows that a suggestion reward system encourages suggestions.

Favourable group dynamics

Knowledge sharing in order to elicit tacit knowledge

Nonaka and Takeuchi (1995) describe the way in which new knowledge is created, as a process involving five phases: (1) sharing of tacit knowledge, (2) creating concepts, (3) justifying concepts, (4) building an archetype, and (5) cross-levelling of knowledge.

The organizational knowledge-creation process begins with the sharing of tacit knowledge, which correlates closely to the socialization process of knowledge conversion i.e. tacit to tacit knowledge. In the second phase, tacit knowledge is shared and converted to explicit knowledge to create a new concept which involves the externalisation process of knowledge conversion. The created concept is then subjected to an internal verification system and if the concept is deemed justified then an archetype is built. This may be in the form of a prototype in the case of "hard" product development or "soft" organisational entity, like a new operating mechanism. Various forms of explicit knowledge are combined at this phase. The last phase cross-levelling of knowledge ensures a wide exchange of knowledge both within the organisation and in exchange with its external environment.

In the new organisation knowledge is shared, not hoarded to increase an individual's power. It is shared in order to create something new. The newly developing concept or product has to be justified and shown to work. Rapid prototypes are created and knowledge is spread throughout the whole organisation.

The system described by Nonaka and Takeuchi (1995) is based in the work-group, and depends for its effectiveness on the quality of the group dynamics within the group. Ideas are delicate and can be most easily destroyed as we have seen for example from the experiments in brainstorming. Brainstorming is a technique of creative problem solving which separates the process of idea generation, from the process of evaluation. Evaluation of ideas is known to diminish the flow of ideas. Ideas can be fostered and developed too, by the use of systems such as brainstorming and Synectics. According to Prince (1980) who developed 'Synectics', the six cognitive processes which take place in idea generation are: wishing, retrieving, imaging, comparing, transforming and storing. In 'wishing' we are motivated to have a new idea. Working in a group can help with this by the rivalry for attention that exists in a group, which does not exist when working on ones' own. In 'retrieving' we retrieve information from the memory that might be useful in the generation of this new idea, or in the solution of this particular problem. In 'imaging' we develop mind pictures of possible solutions, we compare these ideas with existing ones, we transform them where necessary, and we store the idea.

Other more general dynamics are also central to creating a work-group dynamic that is favourable to idea sharing and knowledge development. Included here would be: a climate of inter-personal trust in which risks (such as making possible foolish suggestions) can be taken; a feeling of having a 'stake' in the knowledge being developed; and a sense of excitement and curiosity.

Double-loop learning
According to Argyris and Schon (1978), double–loop learning or 'learning how to learn'
involves the establishment of an organisational culture (learning organisation) that permits

more openness, sharing of feelings, perceptions and assumptions (but also the establishment of an agency that 'supervises' the translation of failures into new ways of working see chapter 2 above.. Organisational norms and constraints are challenged by this process (for more detail see section 1 above-Learning Culture, also see chapter 2 of this book).

Active idea generation and processing
It is a characteristic of much developmental research that it explores a topic in great depth. This deep exploration sometimes has the unintended consequence of making the topic sterile, and the project runs out of creative ideas. This is a well-known psychological phenomenon, which has been recognised by the development of a number of group problem solving techniques, which use both the rational and intuitive approaches to problems.

According to Prince (1980; see no. 6 above), this process of idea generation is greatly interfered with by other learned cognitive processes. These processes include our general habit of evaluating our own performance by existing criteria. This practice tends to diminish our creative capacity because our habits of social thinking, developed in order to help us get on with others, also interfere with our uniquely individual approach to ideas.

Prince lists a number of characteristics of both routine and speculative thinking to illustrate our frequent reluctance to be speculative:

Table 16.2: Characteristics of routine and creative thinking. Prince (1980)

Characteristics of routine thinking:	**Characteristics of creative thinking:**
Logical	You do not know where you are going
Empirical	You do not know how you are going to get
Socially acceptable	there
Few mistakes tolerated	Focus is on the process as well as the result
Focus on task completion	Many mistakes are necessary
Predictable	Much confusion
Comfortable	Much uncertainty High risk
Low risk	Not provable in advance
Socially acceptable	Unpredictable
Supported	Appears inefficient and wasteful
	Makes you anxious
	Easy to reject as impractical or impossible

This impressive list illustrates the difficulty inherent in moving into the truly speculative area of ideas, compared to the more predictable and less risky option of routine thinking. Of course there is an irony in all of this: the more routine the thinking the less creative will be the research, and hence the whole project, and the reputations of those involved, will be at risk. Whereas the more speculative and creative the project is the more it is unlikely to succeed, but if it does do so, it may develop some new insight, and enhance the reputation of those involved!

The decision to take a higher risk direction can be assisted by the group, by the provision of group support and discussion, and by processes for the development and enhancement of ideas. De Bono favours rather similar processes to Prince in order to stimulate his 'lateral thinking' –a mode of thinking most probably associated with discovery oriented thinking. He points out that for effective brainstorming sessions to work, they need the cross-stimulation of other members of a group. In other words the discovery process in science is likely to be helped considerably by working in a stimulating research group, rather than entirely on ones' own (De Bono, 1970). Both techniques, Synectics and Lateral thinking seem to operate by opening up the mind to an increased number of associations and analogies which are useful in the discovery process. The associated group dynamics involved in these techniques can develop the ability to draw analogies, and bring them in from a wide variety of sources. Given that the scientist's mind is already full with the problem being addressed, these techniques can be very useful in reaching novel and imaginative solutions and re-formulations.

Smart foraging
Perkins model, which he calls "smart foraging", fits the process of innovation and discovery very well. The researcher is "well tuned to the topography of ideas" in his chosen general area of research (Perkins, 1981). This model builds on Newell and Simon's (1972) model, which suggest the notion of the 'problem space' as the cognitive area within which the problem is identified. This 'problem space' is the area of the world of knowledge which the scientist or developer is familiar with, which may include a number of disparate areas of research, and which contains problems which are known to him or her. The aim is to arrive at a new target state which will provide an advance on the existing state of knowledge or insight in the defined 'problem space'. With problems that lack a clear specifiable path toward a solution, individuals often have to go beyond and outside the problem space in order to reach towards solutions. As this involves new insight, this idea of problem spaces seems to connect very well with the idea that substantial innovators need freedom to reach new insights in solving problems.

It is clear that methods of developing and pursuing ideas are many and varied, and free open methods need to be available to those talented enough to wish to expand our knowledge.

Avoidance of group think

Janis (1971) describes 'group think' as the mode of thinking that persons engage in when concurrence seeking becomes so dominant in a cohesive in-group, that it tends to override realistic appraisal of alternative causes of action' (Janis and Mann, 1977).

Symptoms of group think include:

1.   The group shares an illusion of invulnerability
2.   The group engages in collective rationalisation to discount dissonant information
3.   The group believes in the inherent morality of what it has to do
4.   The group develops stereotypes of other groups and of dissenters which protects it from accurate analysis
5.   The group puts direct pressure on dissenters in order to silence them
6.   The group censors its own thoughts and doubts
7.   The group believes in unanimity due to lack of dissent " silence = consent'
8.   Some members of the group come to function in the role of mind guards watchmen who protect the leaders from dissenting views by discouraging these views.

If a leader or manager is to prevent these circumstances, conditions must be established that encourage dissent, critical judgement and checking of assumptions. If leaders are not prepared to hear group dissent, then group process need not be used. Group think leads to group polarisation, where the group tends to solve a problem or make a decision in one direction ignoring other alternatives. Group think is an extreme form of group polarisation, which is a tendency for groups to arrive at a more extreme (or polarised) decision on some issue than the average of the individual member's decisions (Manstead and Semin, 1988).

Cognitive processes

Problem finding

Great emphasis in the literature has been on active learning processes, and these processes are indeed those used in the development of new knowledge (Argyris and Schon, 1978). But Brugman (1995) has pointed out that problem finding is a neglected area of the research literature, though the importance of this area is growing at present. He suggests a useful strategy for the expansion of competencies in this area. As long ago as 1917, Dewey, (1917, 1933) explored this topic. There followed a long period of great interest in problem-solving, which has led to many useful results (see Getzels and Csikzentmihalyi, 1976; Perkins, 1981; Sternberg, 1984; Gardner, 1984; Runco 1994)

Problem solving
Within any group, dynamics of different kinds emerge (Anderson and West, 1998). Many groups approach the solution of problems in an entirely rational way. They very often find however, that after a while, progress slows down, and the problem seems to become insoluble. In this case group problem solving strategies which combine rational approaches with intuitive ones, can be useful, even essential.

TESTING THE THEORY: MEASURING THE VARIABLES AND TESTING THE HYPOTHESES

Beginning with the outcome measure, we need to quantify what is meant by knowledge development. We are dealing with a difficult field here where almost no variable is directly definable, measurable or quantifiable. Unlike salary costs, expenses, absenteeism and many other variables, the variables involved are only assessable as perceptions of reality, and therefore valid and reliable measures will have to be developed to assess them.

Does any piece of new knowledge, however trivial, constitute the development of knowledge? Perhaps not, but this illustrates the difficulty and it needs to be overcome. In the absence of a truly objective measure of innovative achievement, we might for example develop a scale of knowledge development, which might look like this:

Table 16.3: Knowledge Development Rating Scale

Knowledge Development Rating Scale	
Major Scientific Discovery	20.0
Scientific Discovery	15.0
Significant Innovation	10.0
Innovation	7.5
New Ideas	5.0
Ideas	1.0

A number of independent expert assessors would rate the outcome of a knowledge development activity and an agreed figure would be allocated to each result. In a similar way we need to be able to measure every other aspect of the theory we propose. The extent of the learning culture; the extent of the existence of the requisite knowledge; the extent of the existence of requisite technical support; the extent of the ability to identify the right problems; the extent of the ability to solve those problems; and the extent of an equitable reward system. The collaborative processes involved in group dynamics need to be at least acknowledged while solving and identifying problems, with all efforts made to avoid group-think.

If we are to verify our cognitive-organisational model of the knowledge management process we must then guide the process with a series of hypotheses based on the related literature. These hypotheses might be framed under the two organisational (favourable organisational conditions and favourable group dynamics) and the cognitive processes factors which affect knowledge development and creation.

Favourable organisational conditions

Hypothesis 1: Knowledge development and creation will be more effective to the extent that the organisation has an equitable reward system.

Hypothesis 2: Knowledge development and creation will be more effective to the extent that the organisation has an effective operating learning culture.

Hypothesis 3: Knowledge development and creation will be more effective to the extent that the organisation has access to requisite and timely knowledge.

Hypothesis4: Knowledge development and creation will be more effective to the extent that the organisation has a highly developed technical support system.

Favourable group dynamics

Hypothesis 5: Knowledge development and creation will be more effective to the extent that the organisation shares knowledge.

Hypothesis 6: Knowledge development and creation will be more effective to the extent that the organisation engages in double-loop learning.

Hypothesis 7: Knowledge development and creation will be more effective to the extent that the organisation engages in active idea generation and processing.

Hypothesis 8: Knowledge development and creation will be more effective to the extent that the organisation uses smart foraging.

Hypothesis 9: Knowledge development and creation will be more effective to the extent that the organisation avoids group think.

Cognitive processes

Hypothesis 10: Favourable organisational conditions will increase problem finding.

Hypothesis 11: Favourable organisational conditions will increase problem solving.

Hypothesis12: Favourable group dynamics will increase organisational problem finding.

Hypothesis 13: Favourable group dynamics will increase organisational problem solving

Hypothesis 14: Knowledge development and creation will be more effective to the extent that the organisation has increased problem finding.

Hypothesis 15: Knowledge development and creation will be more effective to the extent that the organisation has increased problem-solving.

When we have made measurable the processes involved in knowledge development, and seen it as the successful outcome of a good knowledge management system, we can test our hypotheses. When this is done, and only then, will we know whether our current conception of knowledge management is really useful, or another fashion to be written off as part of the human learning process.

What is outlined here is a possible explanatory theory, and as Kurt Lewin pointed out "There is nothing as useful as a good theory". A testing procedure is also given which could form the basis of a programme for assessing the theory.

Note: I am indebted to Sharon Ryan one of my doctoral students for her very detailed and helpful suggestions and improvements on earlier drafts of this chapter.

REFERENCES

Amabile, T.M. (1988). A model of creativity and innovation in organisations. In: *Research in Organisational Behaviour* (B.M. Staw, and L. Cummings, eds). JAI Press, London

Anderson, N. R. and M. A. West (1998). Measuring climate for work group innovation: Development and validation of the team climate inventory. *Journal of Organizational Behavior*, **19**(3), 235-258.

Andrews, F. M. and G. Gordon, G. (1970). Social and organizational factors affecting innovation. In: *Research Proceedings of the Annual Convention of the American Psychological Association,* pp.589-590. APA, Washington DC.

Arad, Hanson and Schneider (1997). A framework for the study of relationships between organizational characteristics and organizational innovation. *Journal of Creative Behavior*, **31**(1), 42-58.

Argyris, C. and D. A. Schon, D. A. (1978). Organizational learning. Addison-Wesley, Reading, MA.

Brugman, G.M (1995).The discovery and formulation of problems. *European Education*, **27**(1), p.38.

Chalmers, A.F. (1990). *Science and its fabrication.* Minnesota Press, Minnesota.

de Bono, E. (1971). *Lateral thinking for management.* McGraw-Hill, New York.

Dewey, J. (1917). *Creative intelligence.* Holt, New York.

Dewey, J. (1933). *How we think.* Heath, New York.

Finerty, T. (1997). Integrating Learning and knowledge infrastructure. *Journal of Knowledge Management,* **1**(2), 98-104.

Gardner, H. (1984). *Frames of mind.* Heinemann, London.

Getzels, J. W. and M. Csikzentmihalyi (1975). From problem solving to problem finding. In: *Perspectives in creativity* (I. Taylor and J. W. Getzels, eds.), pp. 90-116. Airline-Atherton, Chicago

Hager, T. (1995). *Force of nature.* Simon and Schuster, New York.

Hurley, J. (1990). The collaborative imperative of new technology organisations. *Irish Journal of Psychology*, **11**(2), 211-220.

Hurley, J. (1997). *Organisation and scientific discovery: A study among science Nobel prizewinners*. John Wiley, New York.

Hurley, J. (2000). Gestión del conocimiento y competitividad en la industria. [Knowledge management and competitiveness in industry]. *Revista de Psicología Social Aplicada*, **10**(3), 5-23.

Janis I.L., (1971). Groupthink. *Psychology Today* (Nov), 43-46.

Janis I.L. and Mann, L. (1977). *Decision making: A psychological analysis of conflict, choice and commitment*. Free Press, New York.

Kanter, Rosabeth and Moss (1999).The enduring skills of change leaders. *Leader to Leader*, **13** (Summer 1999), 15-22.

Manstead, A.S.R. and G. R. Semin (1988). Methodology in social psychology: Turning ideas into actions. In: *Introduction to social psychology: A European perspective* (M. Hewstone, W. Stroebe, J. R. Codol and G.M. Stephenson, eds.). Blackwell., New York

Mouly, V. S. and J. K. Sankaran (1998).The behaviour of Indian R&D project groups: An ethnographic study. *Advances in Qualitative Research*, **1**, 137-160.

Newell, A. and H. Simon (1972). *Human problem solving*. Prentice Hall, New York.

Nickel, T. M. and J. F. Krems . (1998). Fuehrungsverhalten und mitarbeiterkreativitaet: Eine empirische untersuchung zum betrieblichen Vorschlagswesen. *Zeitschrift Fuer Arbeits & Organisations Psychologie*, **42**(1), 27-32.

Nonaka, I. and H. Takeuchi (1995). *The knowledge creating company*. Oxford University Press, New York.

Pelz, D. C. and F. M. Andrews (1976). *Scientists in organisations*. Ann Arbor, MI: Institute for social research, university of Michigan. (originally published by Wiley in 1966).

Perkins, D. N. (1981). *The mind's best work: A new psychology of creative thinking*. Harvard University Press, Cambridge, MA.

Polanyi, M. (1958). *Personal knowledge*. Routledge, London.

Polanyi, M. (1962). The republic of science. *Minerva*, **1**, 54-73.

Polanyi, M. (1966) *The tacit dimension*. Routledge, London.

Prince, G. M. (1980).Creativity and learning as skills not talents. (Reprinted from the June-July 1980 and September-October 1980 issues of the Phillips Exeter Bulletin).

Revans, R., (1983) *The ABC of action learning*. Chartwell-Brant, Kent, UK.

Runco, M. A. (1994). *Problem finding, problem solving, and creativity*. Ablex, Norwood, NJ.

Senge, P. M. (1990). *The fifth discipline*. Doubleday, New York.

Sternberg, R. J.,(Ed.). (1984.). *Mechanisms of cognitive development*. Freeman & Co, New York.

Thompson, P. and C. Warhurst (Eds.). (1998). *Workplaces of the future*. Macmillan, London.

Tovstiga, G. (1999). Profiling the knowledge worker in the knowledge-intensive organization. *Journal of Technology Management*, **18** (5/6), 731-744.

17

MANAGING THE RISK OF TESTING NEW GENE THERAPIES ON HUMANS

Michael Baram[1]

INTRODUCTION

This chapter deals with the process of introducing a new medical technology into society, learning about its risks by experimental application to selected human subjects, and using the knowledge gained to ultimately make the technology reasonably safe for commercialisation and public use. Discussion focuses on gene therapy, a biotechnological advance intended to cure diseases caused by genetic conditions, and moral responsibility for managing its developmental risks. Like a new drug, a gene therapy must be tested on humans in clinical trials to determine if it is safe and effective and thereby suitable for sale and medical use (Fiscus, 2001). To minimize risks to the persons involved, researchers in the United States must (1) reduce uncertainties prior to beginning a trial, (2) conduct the trial in accordance with specific precautionary procedures; and (3) identify and report any harms arising during the trial so that action can be immediately taken to protect persons in the same trial and other trials. Reporting is also intended to trigger analysis of such "adverse events" and dissemination of findings to guide further development of the gene therapy so that it ultimately becomes a safe and effective product. Because persons have suffered harms in several trials, investigations have been conducted and found that some researchers do not follow the precautionary procedures and many do not report adverse events. Discussion of these problems finds

---

[1] Center for Law & Technology, Boston University School of Law, USA

inadequacies in the ethical and legal framework governing such experimentation, and identifies aspects of research culture, which resist compliance with safety requisites. Consideration is then given to whether knowledge gained from managing risk in other technological contexts through application of safety culture and organizational learning concepts can be adapted to improve safety in the gene therapy enterprise and thereby fulfil moral responsibilities.

GENE THERAPY AND CLINICAL TRIALS

The rapid advance of biotechnology is providing new methods and products which have great potential for improving human health, with much emphasis being given to the development of gene therapies. A gene therapy involves laboratory creation of new genetic material (DNA) and its delivery into targeted cells of a person with the intention of curing the person's genetically-based illness. The strategy is to have this infusion of new genetic material repair a mutated gene or inherited genetic condition which is believed to be a contributing factor or root cause of the person's illness, and thereby defeat the illness. Of the numerous illnesses determined to be genetically-based, such as certain types of cancers and immune deficiencies, many are believed to be amenable to gene therapy.

According to researchers, the new genetic material must be effectively delivered, and in most experiments, this is attempted by using a viral vector to carry the new genetic material into the target cells and "infect" the cells with the new material. The vectors are viruses which normally cause infectious disease but which have been altered so that they lack disease causing potential, yet retain their capability to overcome natural defences and infect cells with the new material.

Like a new drug, a gene therapy must be proven sufficiently safe and effective before it can be approved for sale and use in medical practice. In the United States, the federal Food and Drug Administration (FDA) makes this determination and requires that companies, researchers and other applicants seeking approval must provide reliable information and data from two main sources, laboratory tests on animals, and tests on humans done by qualified medical researchers in "clinical trials." The trials must conform to FDA requirements, which are intended to assure scientifically reliable results as well as safeguard the persons being tested, many of whom are in poor health because they suffer from the illness that the gene therapy is intended to cure. Unlike more conventional types of new products, such as a chemical drug, a gene therapy poses many uncertainties beyond the range of current medical and biological knowledge, and thereby presents a much greater challenge for managing risks to the persons involved in a trial.

The United States is the most active arena for gene therapy research and the largest market for medical advances. Hundreds of clinical trials testing a broad range of gene therapies on many thousands of human subjects have been conducted. Under federal law, the trials are subject to FDA regulations, and trials funded by federal grants from the National Institutes of Health (NIH) must be performed in accordance with NIH requirements as well.

ETHICAL AND POLICY PROBLEMS

The main purpose of a clinical trial is to develop reliable data, which can be used by the FDA to determine whether a new product is sufficiently safe and effective for sale and use. Up to four sequential trials of a new gene therapy, conducted at considerable expense over several years, may be needed to develop sufficient data, with the first two trials posing most risk to the humans involved. Although hundreds of trials of gene therapies have been conducted since 1989, no final application for approval of a gene therapy has yet been submitted to the FDA, presumably because the trials did not produce sufficient proof of safety and efficacy.

Because the trials involve experimentation on humans, FDA and NIH adhere to ethical guidances derived from authoritative sources. These include the Nuremburg Code, enacted in the aftermath of Nazi experiments during World War II, with principles calling for avoidance of cruel and random experimentation and acceptance of experiments involving humans if done for beneficent societal purposes and the persons have voluntarily provided their informed consent; and the World Medical Association's Helsinki Declaration, which extends the Nuremburg principles by emphasizing that medical researchers must give precedence to protecting their patients and research subjects from harm. In addition, the authoritative Belmont Report (1979) of a U.S. advisory commission attempts to provide an ethical framework for government agencies and researchers to follow in supporting and conducting behavioural and biomedical research on humans where such research is intended to benefit society but does not necessarily pose the prospect of therapeutic value to the persons tested..

The Belmont Report is of most relevance to modern research in the fields of genetics and biotechnology but undermines the rigidly protective Nuremburg and Helsinki principles. It conveniently provides a utilitarian approach for overcoming ethical barriers to human experimentation. The approach it offers to agencies and researchers in determining, for example, whether informed consent is necessary in certain research, or whether human participants can knowingly be exposed to pain and the likelihood of harm, is to require that trial proponents perform a qualitative balancing of societal interests against individual interests, and decide on the basis of which interests are dominant. Since researchers and the agencies which support them are convinced of the societal merits of their scientific projects, it is not surprising that their qualitative balancing inevitably favours conduct of their intended research at the expense of individual protection (i.e. that potential societal benefit outweighs the few potential harms that human subjects may incur). The Belmont Report has therefore helped to legitimate a research culture in which the safety of humans is readily subordinated to scientific progress, a culture in which self-serving assertions about societal benefit outweigh rigid ethical principles for human protection.

Another deleterious influence on safeguarding human subjects is that the two agencies involved in approving and regulating trials give primacy to goals other than human safety. FDA, a regulatory agency, is primarily concerned that a trial provide good scientific data which it can someday rely on to determine whether a gene therapy should be approved for sale. NIH, an agency which supports scientific progress, is mainly concerned with demonstrating that the research it funds reflects such progress. Thus, human safety in trials is subordinated to other goals by both agencies, and the safety measures they do require are designed to accommodate achieving these other goals.

Other factors have similar effect, including the considerable influence of researchers and company sponsors of gene therapy trials on FDA and NIH. For example, FDA keeps most information about clinical trials confidential, because proponents of the trials claim that such information contains trade secrets, or proprietary information which may someday be patentable. Although NIH proclaims a public disclosure policy, it also keeps certain trial information confidential. Thus, there is little public knowledge or opportunity for expressing humanistic concerns when inappropriate trials are conducted or harms occur, despite experience with other technologies which demonstrates that transparency and accountability are important conditions for effective risk management (Lo, 2000; Lorman, 2001).

RISK MANAGEMENT OF CLINICAL TRIALS

FDA and NIH, along with various advisory committees and other federal agents, have enacted a multitude of rules, guidances, discussion points, and advisories regarding the safety of humans being tested in clinical trials of new gene therapy products. These pronouncements essentially attempt to promote three functions for managing risks :

- Reduction of Uncertainties prior to commencement of a trial. The researcher must establish that human subjects will not be exposed to "unreasonable risk", based on available information;
- Management of the Trial in accordance with generic safety procedures set by the agencies, and specific safety procedures set forth in the trial's research protocol which is developed by the researcher and approved by his or her institutional review board (IRB) before the trial. Thereafter, deviations during the trial must be reviewed and approved by the IRB.
- Learning from Experience to assure that occurrence of adverse events during the trial will trigger protective interventions, and that knowledge gained from analysing such events will be used by the agencies to make other trials safer.

Pre-Trial Reduction of Uncertainties.

The agencies require that researchers present information, usually in the form of animal data and models of toxicity and immunology, which indicates that the human subjects to be involved will not be exposed to "unreasonable risk." A research protocol is then designed setting forth the procedures to be followed during the trial. Some of these procedures are generic, such as securing informed consent, others are trial-specific and address residual uncertainties such as criteria for selecting human subjects and the plan for monitoring their reactions to the tests.

Many uncertainties remain at the start of a trial for several reasons. Data derived from prior testing of animals (with the gene therapy to now be tested on humans) is often highly species-specific and cannot be reliably extrapolated to other species or humans. Predicting whether new genetic materials will interact with a person's genome beyond the intended remediation is virtually impossible. Estimating human immune response to the aggressive viral vectors used to deliver the genetic materials is highly problematic. Nevertheless, such issues have not prevented the approval of many trials. As a result, the pre-trial function of reducing uncertainties is of limited value, causing all the parties involved (researchers, company sponsors, agencies) to rely on subsequent trial management as the primary means for safeguarding subjects against unknown and unknowable perils.

Trial Management.

This function involves managing the trial in accordance with the research protocol and agency requisites. Key feature include: (1) subject selection according to protocol-based criteria designed to exclude persons with special vulnerabilities to the rigors of testing; (2) administration of the gene therapy according to the protocol; (3) monitoring the subjects for responses, including any adverse events as defined by FDA and NIH; (4) intervening to prevent or mitigate such adverse events, and if necessary, to end testing in accordance with "stopping rules" set forth in the protocol; (5) reporting adverse events to the agencies; and (6) assuring institutional oversight of the trial by requiring that researcher deviations from the protocol must have prior approval of the organization's IRB (Institutional Review Board). Since researchers value autonomy, research culture is permissive and resistant to imposed constraints, and securing IRB approvals can be time-consuming and impede progress, trial management requirements are viewed unfavourably and implementation is grudging.

Responsibilities for trial management are borne by the principal investigator who designed the project, usually a medical doctor specializing in research, and by his organization, such as a medical research organization. Because trial management is burdensome and largely administrative, and many research institutions lack sufficient managerial skills, private contractors are often hired to carry out this function. In addition, companies sponsor many trials and become involved to protect their investment: for example by insisting that certain information be kept confidential because of its potential value for competitive advantage or securing patents. Thus, restrictions may be imposed on researcher publication, presentations at conferences, and other communications, which would disclose information derived from the trial.

Since many trials are subject to both FDA regulation and NIH research grant provisions, confusion arises when the agencies differ. For example, both agencies require that "adverse events" be reported to them but differ in defining such incidents and how the reports will be dealt with. FDA defines a reportable adverse event as any adverse experience involving one of the persons being tested which is "serious and unexpected" and associated with use of the test product. "Serious" is defined as an event which causes death, is life-threatening, requires hospitalisation, causes a significant disability or birth defect, or which requires medical intervention to prevent such outcomes; and "unexpected" is defined as an event which is inconsistent with the risk information presented before the trial in the researcher's investigational plan. Thus, an event that the researcher judges as being caused by a participant's underlying medical condition or something other than the gene therapy and vector being tested in the trial, or as not being a "serious" or "unexpected" adverse event, will often not be reported to the FDA. However, NIH requires that a broader range of serious events is reportable, irrespective of whether they are caused by the product or whether they are unexpected. The NIH reporting procedure is burdensome and resisted by researchers because NIH policy is to publicly disclose reported information, unlike FDA which keeps such reports confidential in their entirety. Researchers dislike the NIH policy because disclosure may cause shame, blame, and other adversities.

Learning from Experience.

As discussed above, researchers are required to identify and report adverse events to the agencies. Obviously, one purpose of this requisite is to immediately trigger medical intervention to help the person who experienced the event in accordance with ethical and humanitarian principles. It is also intended to stimulate learning about the tested product's risks and to disseminate this knowledge among other researchers so that clinical trials of the same product and other trials of similar nature will be halted or made safer. Learning the root cause and contributing factors involved in creating an adverse event and disseminating this knowledge can therefore enable prevention of further adverse events and help researchers proposing new trials to more effectively reduce uncertainties and improve safety for the human subjects to be involved. The knowledge can be further used to adjust the course of research and enable more efficient development of safe gene therapies.

For each of the several hundred trials funded by NIH, the learning process, from reporting to analysis to dissemination, takes place within a structure devised by NIH. Since NIH is a research organization, it lacks capability to effectively monitor and manage these processes and serious deficiencies have been discovered, particularly the failure of researchers to report adverse events, which is likely due to NIH policy of publicizing the events and any useful knowledge gained. Not much is known about adverse event reports to FDA because the agency treats these and virtually all other trial-related information as confidential, causing the learning and knowledge gained to remain with the agency and the researcher who reported the event. Thus, neither agency fulfils the larger purposes of adverse event reporting.

BREAKDOWN AND REPAIRS OF THE RISK MANAGEMENT SYSTEM

The first public realization that the NIH and FDA risk management systems are not sufficiently protective was the news report of the death of 18 year old Jesse Gelsinger, who had been enrolled in a gene therapy trial at the University of Pennsylvania. Gelsinger went into immune shock shortly after being treated with a viral vector bearing a gene therapy product, and died within three days.

This widely publicized incident prompted much public concern and caused FDA and NIH to investigate the death and the conduct of the trial. The agencies established that the test procedure was the definite cause of the death because the immune shock was a reaction to the virus used to carry the genetic material. They also found that the principal researcher, a prominent geneticist, had not complied with many aspects of the research protocol in conducting the trial, including procedures for subject selection, testing, and monitoring, and had further avoided informing his institution's review board (IRB) about his deviations from the protocol. According to FDA and NIH requirements, the IRB must be consulted for review and approval of any deviations from the protocol during conduct of a trial. They also found that Dr. James Wilson, the researcher, held substantial financial interests in the outcome of the trial which had been approved by the university, including the right to secure patents on the knowledge acquired from the trial for his personal gain, and his ownership of a major share of the company which made the gene therapy product he had been testing on Gelsinger and other participants. Disclosure of these financial conflicts of interest shocked public confidence in medical research and led to severe criticism of the agencies, the researcher, and the University.

Put on the defensive, the agencies investigated a large sample of other gene therapy trials and found many researcher violations of protocols as well as inadequate oversight and supervision by the IRB's involved. They also found a high toll of adverse events which included eight deaths and some 700 other harms, none of which had been reported to NIH and only a few of which had been reported to FDA. When confronted, many researchers argued that the events were not reported because they were likely due to causes other than the products they were testing, or were not "unexpected". Another important finding was that many researchers directing clinical trials have ownership interests similar to those of Dr. Wilson, causing concern that financial interests of researchers may be impairing objectivity and data reliability, as well as undermining procedures for human safety and reporting adverse events. Another possible cause of widespread non-reporting, noted earlier, is fear among researchers that NIH's policy of publicly disclosing information on adverse events causes adverse publicity which can lead to blame, potential liability, and other harms to their self-interests.

Thus, evidence has accumulated that many researchers fail to comply with safety requisites, that IRB's, composed mainly of the researcher's busy colleagues, are ineffective overseers, that financial interests of sponsors and researchers may be undermining safety and the credibility of trial results, and that adverse event reporting and learning processes are far more problematic than envisioned. As a result, the agencies, needing human data but under considerable pressure to make trials safer, have promised to repair the safety system.

According to FDA and NIH, several types of repairs are being made. New programs and guidances (guidelines?) have been developed to educate researchers and IRB's on ethical responsibilities and regulatory requirements. More investigations, stringent enforcement and severe monetary penalties have been promised.. Medical research institutions have been told to prevent egregious conflicts of interest, and to consider having independent third parties observe trial management. National associations of medical institutions have been urged to develop a system for certifying qualified research groups. FDA and NIH have agreed to share adverse event information, and to develop standards for the data and models to be used to prove "no unreasonable risk" when approval is sought for future trials.

While these and other repairs are being made, trials are being approved and carried out until adverse events occur, at which time they are halted by the agencies, the events examined, and decisions then made as to whether to resume. In some reported cases, decisions to resume trials are being made without anyone having learned how to prevent replication of the adverse events which caused the suspensions. Instead, trials are being resumed when it has been determined that persons suffering the disease at issue have no other therapeutic options, that the healing potential of the gene therapy outweighs the risk to such persons, and that each participant has consented after being fully informed of the adverse events and other known risks. Thus, not much has really changed and safety continues to be subordinated to scientific advance and other research interests, a morally troubling situation.

REAL REFORM

So long as there is support for the proposition that gene therapy will lead to remarkable new methods of curing disease, and so long as animal testing and other methods for assessing risk are unreliable, human test data will be needed to determine whether a gene therapy is safe and effective. Thus the need for clinical trials remains. But so does moral responsibility to use best efforts and practices to protect the thousands of participants in these trials and to progressively reduce risks over time. The existing array of safety measures, many of which were hasty responses to adverse events and public criticism, do not meet this standard. They fail to address deeply imbedded values, norms and behavioural aspects of biotech research culture, which are known to have the persistent effect of subordinating safety to the interests of researchers, sponsors and agencies. Thus, a deeper consideration of how risks should be managed in the development of gene therapies is needed. In this undertaking, it may be of high value to consider best practices developed in other technological contexts and whether they can be appropriately adapted to the clinical trials enterprise.

The preceding discussion indicates that a morally responsible approach to managing risk under conditions of great technical uncertainty in the gene therapy context requires attention to two core issues:

- the deeply imbedded values and motives of the parties involved (researchers, medical institutions, corporate sponsors) which subordinate safety to their financial and promotional interests and thereby undermine performance of safety and reporting procedures, and
- the fears of these same parties that reporting adverse events for evaluation and disseminating the knowledge gained will cause them to incur blame, liability, loss of competitive advantage, and other adversities.

Thus, priority should be given to searching for best practices, which have directly addressed similar issues in other technological contexts. For example, the growing acceptance and implementation of the "safety culture" concept in the nuclear power context, and efforts to promote organizational learning from accidents and near misses in several technological sectors (e.g. the chemical process industry, air and rail transport systems) are worthy of study.

These other technological contexts obviously differ from the gene therapy trial context in many respects: e.g. the scale of the harms to be prevented, the degree of uncertainty about the technology and how to use it safely, the extent to which the persons at risk are willing participants. But the issues being addressed are fundamentally the same as the core issues in gene therapy trials, namely, how to mitigate individual and organizational values, motives, beliefs and consequent behaviours which evince disregard for safety and create resistance to prescribed safety measures; how to elevate safety and make it an integral part of the overall system for productively managing the technology; and how to overcome reluctance to report adverse events in order to improve learning and knowledge sharing (Baram and Proctor, 1993; Baram, 2001).

Safety culture

Safety culture is an aspirational concept that was formulated, following Chernobyl, to encourage improvements in the management of nuclear power facilities for the purpose of preventing catastrophic accidents. As illuminated by Wilpert and Itoigawa (2001), the concept is based on recognition that imposing detailed safety rules and procedures and requiring rote compliance is insufficient because safety is a socio-technical matter and depends on interactions of human and technical factors in an organization. Thus, a safety culture is needed to assure that unconscious assumptions, values, norms and consequent behaviours at each facility do not impede performance of requisite safety procedures.

Work by other analysts has reinforced the safety culture concept and further defined its purposes. For example, Rasmussen and Svedung (2000) have emphasized the organizational need to prevent subtle erosion of defences against accidents by being attentive to the behaviour of decision-makers within the organization and associated external parties (such as regulators) and the information they exchange. Vaughn (1996)has warned against the normalization of deviance, which subtly occurs within organizations over time, usually in response to competitive pressures and production goals. Turner and Pidgeon (1997), Wilpert (2001) and others have emphasized that a chief characteristic of safety culture is the fostering of a robust learning system, one which encourages continual reflection upon organizational practices through monitoring, analysis and feedback, and encompasses all organizational personnel, external regulators and other stakeholders.

Other important characteristics of safety culture have been identified and promoted by nuclear power regulators. As summarized by Meshkati (1999), these include "a prevailing condition in which each employee is always focused on improving safety, is aware of what can go wrong, feels personally accountable for safe operation, and takes pride and ownership in the plant", and "an insistence on a sound technical basis for actions and a rigorous self-assessment of problems". However, proponents of safety culture caution against reducing the concept to practice manuals and checklist procedures because they rob the concept of its vitality and dynamic qualities.

As a result, safety culture remains a qualitative and aspirational concept which differs in its application at each facility, and its accomplishments may not be measurable. But its influence is pervasive because it has raised the generally accepted standard of care for preventing nuclear accidents by introducing human factors into what had been a technocratic, production-driven management system, by calling for continuous improvement through organizational learning processes which engage all personnel, and by legitimating continuous attention to safety as a behavioural norm to the extent that those who wish to deviate by subordinating safety have a greater burden of justifying their action.

Bringing safety culture into the gene therapy context would involve delineating how it would address the special risk circumstances of clinical trials. For example, it would encompass many parties (e.g. researchers, IRB's, human subjects, medical research institutions, corporate sponsors and contractors); promote deliberations between researchers and IRB's about precautionary measures for dealing with residual technical uncertainties (in addition to stressing adherence to required procedures); and subordinate the special interests of the parties (e.g. financial, confidentiality, etc.) to the main goals of the trial, namely its safe conduct and generation of scientifically reliable (objective) data. Perhaps the deepest reform could be accomplished by infusing the safety culture concept into the Belmont Report and thereby produce a revised ethical framework which restores the primacy of safeguarding persons at risk in gene therapy trials and subordinates scientific progress to human safety. Bringing safety culture into the gene therapy context, in which thousands of humans are enlisted and put at risk for potential societal benefit, presents a considerable challenge, which leading researchers and agencies should undertake as a moral responsibility.

Learning from adverse events

Similarly, the practice of organizational learning from incidents of experience in order to prevent future harm, now being applied in other technological contexts, needs to be considered to remedy inadequacies in the learning system devised for gene therapy trials. Reduced to fundamentals, an organization's system for learning from incidents of experience must have the following generic features: an overall concept of what knowledge is to be sought from operating experience (goals), a definition of the incidents or aspects of experience that when evaluated are likely to produce the desired learning (such as accidents, near misses, errors, etc.), a design of the learning system which makes it an integral part of operations and capable of capturing the incidents in timely fashion, and designation of responsibilities and procedures for reporting and evaluating the incidents and thereafter sharing the knowledge gained and assuring it is put to use.

Analysts of such learning systems in various technological risk contexts have dealt with these features and the problems encountered. For example, Freitag and Hale (1997) have discussed three potential goals and the "learning loops" involved: one loop which enables immediate corrective action to prevent an imminent harm, another which enables planning and making deeper changes in operations and management tactics in order to avoid replication of the incident, and a third loop which enables redesigning the operational system and reforming organizational culture and strategies in order to eliminate any residual risk of replication. Baram (1997) and Rosenthal (1997) have dealt with the difficulties in defining what should be reportable for optimal learning within resource constraints, and impediments to reporting and evaluating incidents, such as shame, blame, liability, work stoppage and costs. Fahlbruch and Wilpert (1997) have suggested criteria for choosing an appropriate approach to incident analysis. And many others, including Tamuz, Carroll *et al.* (2002), Koornneef and Hale (2001), Wright (2001) and Leape have evaluated experience with learning systems in air and rail transport systems, the chemical process and nuclear power industries, and the hospital context.

For example, Wright has evaluated the Confidential Incident Reporting and Analysis System (CIRAS) used to improve safety among all companies in the U. K. rail system. Her study indicates the value of having an independent body administer CIRAS because this mitigates the competitive pressures which would otherwise impede incident reporting and analysis functions in each company. Further, it has enabled creation of an industry-wide data base and facilitated knowledge dissemination among all companies without raising concerns about blame and liability.

Similarly, Koornneef has evaluated hospital units which use medical devices and found the need to "anchor organizational learning firmly at the level of the primary processes of the organization so that it involves and rewards the efforts of people working there". This would assure that they "see the value of the learning process" and transform the knowledge gained into action. In addition to single loop learning to modify tasks, he suggests additional learning loops to address the need to change norms and values, and the development of accessible databases for knowledge sharing purposes. Given the fragmentation of activities and time pressures on personnel in the hospital environment, he concludes with the recommendation that the learning process for safe medical device use, whether it comprises one entity or multiple entities and regulators, needs a dedicated learning agent or agency to coordinate and manage the learning process.

Such studies provide useful insights for refining and improving learning processes. But they may also lead to problematic developments. For example, electronic methods of knowledge management are now being deployed to reduce medical errors in hospitals. Because a leading physician "needs to know something about 10,000 different diseases and syndromes, 3,000 medications, 1,100 laboratory tests, and many of the 400,000 articles added each year",

Davenport and Glaser (2002) find that knowledge management for error reduction needs more than employee networks and communities of practice. Thus, they recommend that hospitals "bake specialized knowledge into the jobs of highly skilled workers-to make the knowledge so readily accessible that it can't be avoided", "embed it into the technology that knowledge workers use to do their jobs" and accomplish this by "linking massive amounts of constantly updated clinical knowledge to the information technology systems that support doctors' work processes."

As they recognize, a "real time" system for learning and managing knowledge can overload intended users, a situation which prompts technical assistance. But if such assistance leads to technological solutions, which eliminate or marginalize the need to be attentive to individual values, cognitive processes, and behaviour, a familiar problem arises. As noted earlier, the technological approach to safety proved to be inadequate in the nuclear power domain because it precluded psychological and behavioural considerations. It is therefore essential that learning and knowledge management systems be deployed in conjunction with a strong safety culture program to prevent such an outcome.

Although knowledge about the efficacy of learning systems elsewhere may be of value for managing risk in the gene therapy context, deriving this value will depend on identifying distinguishing features and needs of the gene therapy context and adapting the experiential knowledge accordingly. For example, the learning systems used elsewhere are usually aimed at preventing large scale accidents in managing a technology which poses few uncertainties, focus on near misses and prior accidents to discern errors of commission or deviations from standard safety procedures (e.g. worker mistakes), and seek to prevent recurrences of incidents by correcting the errors and deviations without changing the technology or addressing its organizational context and culture. However, in the gene therapy context, the learning system needs to aim at preventing individual harm where each individual has unique vulnerability characteristics and the technology being used presents great uncertainty as to how to manage it safely. Further, it must focus on discerning latent hazards as well as deviations from incomplete and untested safety procedures, and seek to prevent incidents which have not occurred previously by changing the technology when merely correcting deviations seems insufficient. Finally, the gene therapy enterprise requires attention to double or triple loop learning and the employment of learning agents in costly and technically complex incident analysis processes in order to minimize risks in current and future trials and facilitate progress towards development of safe gene therapies. Thus, the challenge presented by gene therapy is considerable but essential from a moral perspective.

CONCLUSION

The gene therapy enterprise has potential for providing health benefits of incalculable value. However, of necessity, it involves experimentation on thousands of human subjects, persons who are exposed to health risks in clinical trials. The current system for minimizing these risks and managing knowledge about these risks is inadequate. It enables researchers, sponsors and regulators to subordinate human safety to scientific progress and other interests, and does not address the values, norms and behaviours which resist existing safeguards.

Thus, intrinsic to the gene therapy enterprise is the problem of managing risk in a morally responsible manner, a problem which deserves, at the least, the application of best efforts and best practices under conditions of great technical uncertainty. To meet this challenge, it will be necessary to learn from advances in safety science, such as concepts and models of safety culture, organizational learning and knowledge management, and their application in other technological domains.

REFERENCES

Baram, M. and S. Proctor (1993). Human gene therapy research: Technological temptations and social control. *The Genetic Resource,* **7** (10).

Baram, M. (1997). Shame, blame and liability: Why safety management suffers organizational learning disabilities. In: *After the event* (A. Hale, B. Wilpert and M. Freitag, eds.), Pergamon, Oxford.

Baram, M. (2001). Making clinical trials safer for human subjects, *American Journal of Law & Medicine,* **27** (2 & 3).

Carroll, J., J. Randolph and S. Hatakenaka (2002). The difficult handover from incident investigation to implementation. In: *System safety* (B. Wilpert and B. Fahlbruch, eds.). Pergamon, Oxford.

Davenport, T. and J. Glaser (2002). Just in time delivery comes to knowledge management. *Harvard Business Review,* **7**.

Fahlbruch, B. and B. Wilpert (1997). Event Analysis as Problem Solving Process. In: *After the event* (A. Hale, B. Wilpert and M. Freitag, eds.). Pergamon, Oxford.

Fiscus, P. (2001). Protecting Human Research Subjects. *Journal of BioLaw & Business,* **4**(4).

Food and Drug Administration, U. S. Government: Regulations, 21 *Code of Federal Regulations* sections 46, 56, 76: and *Guidance for Human Somatic Cell Therapy and Gene Therapy.*

Freitag, M. and Hale A. (1997). Structure of event analysis. In: *After the event* (A. Hale, B. Wilpert and M. Freitag, eds.). Pergamon, Oxford.

Koornneef, F. and A. R. Hale (2001). Organisational memory from Operational Surprises: Some Ins and Outs, *presentation at Bad Homburg Conference.*

Lo, B. (2000). Conflict of Interest Policies for Investigators in Clinical Trials, *New England Journal of Medicine,* pp.1621 et seq.

Lorman, A. (2001). Clinical Trials Face Heightened Scrutiny, *Journal of BioLaw & Business,* **4**(4).

Meshkati, N. (1999). The cultural context of nuclear safety culture. In: *Nuclear safety: A human factors perspective* (J. Misumi, B. Wilpert and R. Miller, eds.). Taylor & Francis, London.

Misumi, J., B. Wilpert and R. Miller (eds) (1999). *Nuclear safety: A human factors perspective.* Taylor & Francis, London.

National Commission for the Protection of Human Subjects of Biomedical and Behavioral Research, (1979). *Ethical Principles and Guidelines for the Protection of Human Subjects of Research* (the Belmont Report), U.S. Government.

National Institutes of Health, *Guidelines for Research Involving Recombinant DNA Molecules, Appendix M*, U. S. Government.

Rasmussen, J. and I. Svedung (2000). *Proactive risk management in a dynamic society.* Swedish Rescue Services Agency.

Rosenthal, I. (1997). Major event analysis in the U. S. chemical industry. In: *After the event* (A. Hale, B. Wilpert and M. Freitag, eds.). Pergamon, London.

Trials of War Criminals before the Nuremberg Military Tribunal, V. I & II, The Medical Case, (1948). U.S. Government.

Turner, B.and N. Pidgeon (1997). *Man-made disasters.* Butterworth

Vaughn, D. (1996). *The Challenger launch decision.* Univ. Chicago Press.

Wilpert, B. (2001). The relevance of safety culture for nuclear power operations. In: *Safety culture in nuclear power operations* (B. Wilpert and N. Itoigawa, eds.).Taylor & Francis, London.

Wilpert, B. and N. Itoigawa (eds.) (2001). *Safety culture in nuclear power operations.* Taylor & Francis, London.

World Medical Association (1964, amended 1996). *Declaration of Helsinki.*

Wright, L. (2001). Information collection and knowledge transfer in the UK Railways via CIRAS, *presentation at Bad Homburg Conference.*

106th U. S. Congress (2000). *Gene Therapy Hearing: Is There Oversight for Patient Safety?*

18

ORGANISATIONAL LEARNING: WHAT HAVE WE LEARNED

J. H. Erik Andriessen[1] and Babette Fahlbruch[2]

This is a book about organisational surprises – of two types – and how individuals and organisations can learn from them. The concept of organisational surprises is coined to cover two phenomena that at first sight seem to be quite divergent, i.e. accidents and new experiences. Smart individuals and organisations want to learn from these surprises, either to prevent them next time or to use these new experiences for better products and services. The question is how this learning can be optimised. How should is knowledge transfer be organised in such a way that in actual practice maximal use of existing knowledge is made, either to solve existing problems or for the development of innovative approaches.

In this book several examples are presented of knowledge transfer in organisations, on the one hand via storage in and retrieval from information systems, and on the other hand via interpersonal communication, i.e. via meetings and committees or communities of practice. Interpersonal communication seems to be quite successful but probably not under all circumstances. And information systems, both in the high risk sectors and in the service sector, are often not used (well) according Kjellén, Wahlström, and Huysman, in this volume). Kjellén concludes: *The results of the evaluation paint a rather discouraging picture of the possibilities of capturing individual department members' knowledge and to make it explicitly available as*

[1] Delft University of Technology, The Netherlands
[2] Berlin University of Technology, Germany

a database of Best practice to the department as a whole. Other systems, however, appear to be used quite satisfactorily (Wright, Dhondt).

What are the pros and cons of the various approaches, why is learning in some cases more successful than in others, and particularly, can the one sector learn from the lessons in the other? From the illustrations in the case based chapters and the theories in the conceptual chapters we try to develop answers to these questions.

THEORY

Whether we talk about learning from incidents or learning from successes, it is all about learning, by individuals and by organisations. We are in fact talking about on the one hand learning *processes* and on the other about repositories or *memories*. Can organisations learn, like individuals? Do organisations have a memory, like individuals have? Lehner warns ' ... *the term "organizational memory" should in no way be considered analogous to a "brain" to which organizations have access. The term is simply meant to imply that the organization's employees, written records, or data "contain" knowledge that is readily accessible'*. However, organisational memory is only effective when it is considered in the same way as individual memory. Human remembering is not a simple picking of rigid data from a fixed repository, but an active process whereby memory elements data are reprocessed in relation with the actual setting. In the same way organisational memory works only effectively when it can be contextualised and adapted to the present setting (Bannon and Kuutti, 1996).

Knowledge management concerns the questions: How should (individual and organisational) learning be organised and how should organisational memories be given a role in these processes. Both for individuals (Sonnentag) and for organisations proper 'knowledge management' helps to overcome problems of overload, chaos and reinventing the wheel.

In the past decades too much emphasis has been laid on the 'codification' strategies (see chapter 1) and not much on the personalisation strategy. However, the best learning at the individual level appears to be 'situated learning' (Lave and Wenger, 1991). The rationale behind this view is the fact that knowledge is different from simple information. It is information that is experienced and interpreted by a person, it is related to an actual situation and it makes sense to that person to the extent that it is seen as relevant to actual problems. Just reading solutions in a book or a database is seldom effective, because *"...much knowledge – in the sense of personal experiences – is very difficult to explicate, and secondly because of the psychological resistance against providing and against using knowledge that is separated from its owner, i.e. that is made impersonal ... Exchanging knowledge with others may provide status and is trustworthy for the receiver. Putting knowledge in a system is cumbersome, removes it from its context and rarely provides personal rewards"* (Andriessen *et al.*).

Organisational learning means that the solutions individuals have found for their problems is somehow shared and transferred to the organisation. This implies that it should be an interaction of processes on three levels: individual, group and organisation (Huysman). And in the case of product-related incidents, even the supplier and regulator level should be involved (Baram). The question is how to transfer this individual learning. Following Argyris and others (see Koornneef) this implies single and double (and sometimes triple) loop learning. Koornneef and Hale write: *'There is a fundamental problem inherent in the notification process which is that the notifying message becomes divorced from its context in the process of communication Therefore, the learning agency has a role in recapturing and preserving the contextual information lost in the notification process'.* According to Koornneef and Hale, and also Dhondt *et al.*, there has to be a learning agency to ensure that the learning experience becomes embedded in the organization. The learning agency should have the task of making additional inquiries if the notification is not enough to act as the basis for a decision"

Four different transfer systems can be distinguished, i.e. Nonaka's internalisation, socialisation externalisation and combination (see Dhondt):
- Socialization of knowledge: from 'tacit' to 'tacit'. Examples of this type of transfer are direct interaction in e.g. communities of practice, customer and supplier contacts;
- Externalisation of knowledge: from 'tacit' to 'explicit'. Examples of this type are writing down of events and storing them in a database; formal presentations to peers; providing documents with best practices
- Combination of knowledge: from 'explicit' to 'explicit'. Examples of this type are combining different written or electronic sources into a new one
- Internalisation of knowledge: from 'explicit' to 'tacit'. Examples of this type are learning by doing, reading, courses

A smart combination of these systems is the optimal way of organisational learning. In such context knowledge systems can also be used fruitfully, particularly the latter generations of such systems, providing for communication between stakeholders (Maier-Lehner).

Interesting is the finding – reported on by Kjellén - that the very process of **developing** such knowledge system appears to be more effective than the existence of the system once it is ready. At first sight this seems to be remarkable, but it fits in well with the theory presented above. Working on such a system together implies that the group of people involved, develop common visions and meanings, based on their common context. However, once the system was ready, new users did not share the same background and the opportunity to discuss certain data, and therefore did not make much use of it.

So, the optimal process of organisational learning appears to require a combination of interpersonal transfer and codification. This is e.g. illustrated by some case studies on communities of practice (Andriessen), Wahlström's study on a quality system, and particularly Wright's study on the use of the incident reporting system CIRAS.

CIRAS is the '*Confidential Incident Reporting and Analysis System*' currently being used in order to identify and deal with human factors problems on the railways in the UK. A characteristic of CIRAS is that it is confidential and administered by an independent body. But a major strength is the fact that the data are not only analysed carefully and reported in journals and newsletters, but also that they are collected and disseminated face to face in various groups and contacts. Information on accidents, incidents and situations is collected via a report form, followed up by a telephone or face to face interview. A full interview is found to be absolutely essential as railway staff often write only minimal details on the reporting form.

CIRAS staff extricate valuable information from the interview and pass this to individual company representatives. The information is turned into a situated contextual based narrative detailing the problem. Company representatives and the CIRAS staff discuss these outlines of events individually. Moreover, in the Liaison Committee meetings the company representatives get together with CIRAS to discuss the reports from their various perspectives. The groups operate at a regional level, which provides for optimal contextualising of the discussions.

CONDITIONS

Optimal transfer of knowledge according to the lines indicated above is not a simple affair. It requires quite some effort and creativity and it requires favourable circumstances. The authors in this book have discussed various conditions that have to be realised for knowledge transfer and learning to be possible and effective.

Several authors refer to the fact that *a general climate of co-operation and trust* within an organization has bearings on the members' willingness to share experience. Kjellén reports that in his case organisation a considerable decrease in these factors took place between 1998 and 2000. The single most significant explanation for this decrease was the merger and re-organization carried out about half a year before the survey.

Huysman places this climate of trust in the wider concept of 'social capital', which refers to social networks that create opportunities to connect. Social capital has three dimensions that are highly interrelated (Nahapiet and Goshal, 1998), i.e. the earlier mentioned relational dimension of mutual trust, but also a cognitive dimension such as shared codes and language and a structural dimension such as network ties and configurations. Investing in social capital seems particularly important with the growth of virtual organizations, where social networks based on reciprocity, trust and mutual appreciation will last much longer than engineered networks such as organizational teams.

The contributions of Wright and of Dhondt *et al.*, however, show that organisational and psychological factors are to some extent contrary to full scale cooperation and sharing. As Dhondt *et al.* point out the context in organisations is generally not (only) one of trust and openness. Knowledge transfer mechanisms within companies are linked to power relationships and organisational types. Simple bureaucracies – where control is generally very centralised – are less conducive to knowledge sharing than adhocracies (see Mintzberg, 1983), so the implementation of knowledge management must take into account the type of organisation. Moreover, the case studies discussed show that many managers are reorganising companies and not much interested in exploiting hidden knowledge.

Wright's case study of the CIRAS incident system confirms again that sharing incidents may be threatening to people's reputation and position. It will therefore only take place under clear conditions of confidentiality and security. Sharing successes, however, at least in interpersonal communicative settings, can be rewarding in various aspects, such as enhancing reputation.

The final question than is, whether it is possible to promote and facilitate knowledge sharing by concrete measures. The contributions in this book show the positive effects of the various kinds of organisational arrangements.

Lehner, Hurley and Koornneef and Hale point at the role certain knowledge systems (including incident notification systems) can play in supporting interpersonal knowledge sharing processes. According to Maier and Lehner particularly the advanced types of 'interactive' and 'bridging' knowledge management systems can have this role, but they point to the necessity of ease of use of these systems, and of technical provisions for security and reliability. They also conclude that some kind of management (e.g. information or knowledge desk) is needed to prevent „information overload".

Koornneef and Hale, emphasise the necessity of special learning agencies, which according to Argyris' (1996) learning theory, are responsible for translating local experiences into second loop organisational learning. In the same vein Lehner makes a plea for the creation of new professions, such as change agent, knowledge architect and Organisational Memory administrator. This is comparable to what is called in other organisations 'knowledge brokers'.

Hurley points further at the role of adequate reward systems, such as group reward and suggestion reward systems. He also underlines the importance of favourable group dynamic variables, including the skills for active idea-generation and processing.

Finally, Andriessen *et al.* indicate the important role communities of practice can have in exchanging experiences and developing new knowledge for the organisation. They identify several types of communities and some success conditions for each of them. Although communities of practice are should explicitly NOT be managed as organisational units, they can be facilitated by receiving certain freedoms to operate, adequate reward systems, communication technology and management sponsorship.

In conclusion, it has become clear that knowledge sharing and learning, whether individual or organisational learning, are processes of high promises but hard realisation both in the high risk sector and in the services sectors. We are convinced that both sectors can learn from each other and with this book of readings we have provided a set of very practical learning experiences. Lessons to be learned from the high risk sector could be the way to analyse what happened, i.e. analysing successful projects or services to identify the causes for the success with more standardized systematic methods as well as reporting systems and feedback loops could be a lesson to be learned and might high-lighten the reasons and causes for successes. From the service sector we can learn how important it could be to focus on daily operations and future improvements in a more proactive way as well as how networks and communities of practice can help to improve the knowledge sharing process. Together with the conceptual notions in the earlier and final chapter we trust to have contributed to the further development of theory in the field of organisational learning.

REFERENCES

Argyris, C., and Schön, D. A. (1996). *Organisational Learning: theory, method and practice.* Amsterdam, The Netherlands: Addison Wesley.

Bannon, L. J., and Kuutti, K. (1996). *Shifting Perspectives on Organizational Memory: From Storage to Active Remembering.* Paper presented at the 29th Annual Hawaii International Conference on System Science, Maui, Hawaii.

Lave, J., and Wenger, E. (1991). *Situated Learning. Legitimate Peripheral Participation.* Cambridge: University Press.

Mintzberg, H. (1983). *Structures in fives: designing effective organizations.* Englewood Cliffs: Prentice Hall.

Nahapiet, J., and Ghoshal, S. (1998). Social Capital, Intellectual Capital, and the Organizational Advantage. *Academy of Management Review, 23*(2), 119-157.

Nonaka, I., and Takeuchi, H. (1995). *The knowledge creating company.* Oxford: University press.

INDEX